Research Report

T0127935

Rapid Acquisition of Army Command and Control Systems

Shara Williams, Jeffrey A. Drezner, Megan McKernan, Douglas Shontz, Jerry M. Sollinger

Prepared for the United States Army
Approved for public release; distribution unlimited

RAND ARROYO CENTER

The research described in this report was sponsored by the United States Army under Contract No. W74V8H-06-C-0001.

Library of Congress Cataloging-in-Publication Data is available for this publication.

ISBN 978-0-8330-8510-8

The RAND Corporation is a nonprofit institution that helps improve policy and decisionmaking through research and analysis. RAND's publications do not necessarily reflect the opinions of its research clients and sponsors.

Support RAND—make a tax-deductible charitable contribution at www.rand.org/giving/contribute.html

RAND® is a registered trademark.

RAND OFFICES
SANTA MONICA, CA • WASHINGTON, DC
PITTSBURGH, PA • NEW ORLEANS, LA • JACKSON, MS • BOSTON, MA
CAMBRIDGE, UK • BRUSSELS, BE
www.rand.org

Preface

In the last eight years, the Global War on Terrorism (GWOT, subsequently Overseas Contingency Operation or OCO), Operation Enduring Freedom (OEF), and Operation Iraqi Freedom (OIF) have all presented urgent technological challenges for the U.S. military. These challenges, particularly the use of improvised explosive device (IEDs), have forced the Department of Defense (DoD) to move material solutions through the acquisition lifecycle at unprecedented speed.

This study examines how the Army can better manage systems acquired through nontraditional means (i.e., outside the process defined by DoDI 5000.02), focusing on command and control (C2) systems. The research identifies issues, challenges, and problems associated with nontraditional rapid acquisition processes and recommends ways for the DoD acquisition system to more rapidly develop, procure, and field effective C2 systems within the framework of current policies and processes. The research assesses past and current experience in the rapid acquisition of C2 systems through nontraditional means, including detailed case studies of three such systems. Those three case studies are published separately in a web-only companion document entitled *Rapid Acquisition of Army C2 Systems: Case Studies*, RR-210-A (not available to the general public).

This research was performed over the period June 2009 through September 2010. Most of the data collection for the three case studies, including on-site interviews and review of official program documentation, occurred from August 2009 through May 2010, with some later revisions in the 2012–2013 time period. The main body of work in this document reflects information as of the earlier period. The findings should interest Program Offices, Program Executive Offices within the Army and the other military services, the Office of the Secretary of Defense (OSD), Congress, and contractors with an interest in doing business with DoD.

This research was jointly sponsored by the Director, Acquisition and System Management, Assistant Secretary of the Army for Acquisition, Logistics and Technology (ASA(ALT)); the Deputy Chief of Staff (DCS) Logistics (G-4); and the Director, Space and Missile Defense Battle Laboratory, Army Strategic Command. It was conducted within RAND Arroyo Center's Force Development and Technology Program. RAND Arroyo Center, part of the RAND Corporation, is a federally funded research and development center sponsored by the United States Army.

Questions about this report should be directed to the project lead, Jeffrey A. Drezner (Jeffrey_Drezner@rand.org). Questions concerning RAND Arroyo Center's Force Development and Technology Program should be directed toward its director, Chris Pernin (pernin@rand.org).

The Project Unique Identification Code (PUIC) for the project that produced this document is ASPMO09188.

For more information on RAND Arroyo Center, contact the Director of Operations (telephone 310-393-0411, extension 6419; fax 310-451-6952; email Marcy_Agmon@rand.org), or visit Arroyo's web site at http://www.rand.org/ard.html.

Contents

Preface.. iii

Figures... vii

Tables.. ix

Summary xi

Acknowledgments.. xxiii

Glossary xxv

1. Introduction..1

2. Case Studies 13

3. Urgent Need, Rapid Acquisition, and Transition Processes 39

4. Conclusions and Recommendations 65

Appendix

A. Joint/Army Rapid Capabilities and Materiel Developments Initiatives 75

Bibliography 99

(This page is intentionally blank)

Figures

S.1. Study Motivation xii
S.2. Factors Enabling Rapid Acquisition of Army C2 xiv
S.3. War Drives Many Factors Enabling Rapid Acquisition, and the Factors
      Influence Each Other xv
S.4. Summary of Key Findings on Army Rapid Acquisition of C2 xvi
S.5. Institutionalizing Rapid Acquisition Poses Both Challenges and
      Opportunities xviii
S.6. Recommendations for the Army to Improve Rapid Acquisition of C2 xix
1.1. Rapid Acquisition of Army C2 Systems 1
1.2. The Problem 2
1.3. Rapid Acquisitions of Command and Control Capabilities Face Challenges 4
1.4. Study Objective 6
1.5. Study Approach 7
1.6. The Organization of This Report 11
2.1. The Section on Case Studies 13
2.2. The Three Case Studies Span a Range of Organizations, Funding Sources,
      Program Sizes, and Outcomes 14
2.3. Fire Coordination Cell 16
2.4. FCC History 18
2.5. FCC Lessons Learned 20
2.6. Command Post of the Future (CPOF) 22
2.7. CPOF History 23
2.8. CPOF Lessons Learned 25
2.9. The Joint Network Node (JNN) System 28
2.10. JNN History 29
2.11. JNN Lessons Learned 31
2.12. Some Factors Enabling Success Are Unique to Wartime 33
2.13. War Drives Many Factors Enabling Rapid Acquisition, and the Enabling
      Factors Influence Each Other 35
3.1. Introducing the Survey of Rapid Acquisition Processes 39
3.2. Three Sets of Processes Are of Interest 41
3.3. Process Interactions and Interdependencies Can Be Relatively Complex 44
3.4. Recent DSB and GAO Reports on Urgent Needs Processes Identify Similar
      Sets of Challenges 47
3.5. Urgent Need and Rapid Acquisition of C2 Systems Face Many Challenges 50
3.6. IT System Acquisition Presents Its Own Challenges 52

3.7. Capabilities Development for Rapid Transition (CDRT) Process Addresses Transition Decision 54
3.8. Example Programs (1) 57
3.9. Army's Immediate Warfighter Needs (IWN) Process............................59
3.10. Lessons Learned from the Survey of Established Rapid Acquisition Processes 60
3.11. Institutionalizing Rapid Acquisition Poses Both Challenges and Opportunities...62
4.1. Summary of Key Findings...65
4.2. Summary of Recommendations...69
4.3. Document and Preserve Recent Rapid Acquisition and Transition Experience..72
A.1. Joint Chiefs of Staff and Joint Rapid Acquisition Cell Process for Reviewing, Validating, and Fulfilling JUONs.......................................81
A.2. Rapid Equipping Force's Decision-Making Process 84
A.3. JIEDDO's Capabilities and Acquisition Process....................................86
A.4. Capabilities Development for Rapid Transition Process 89

Tables

A.1. Select Joint/Army Rapid Capabilities and Materiel Developments Initiatives, Processes, and Organizations...76

(This page is intentionally blank.)

Summary

For the past decade, the U.S. Army has been engaged in extended overseas conflicts in Afghanistan and Iraq. These conflicts tested the technologies the military developed during the preceding Cold War and post–Cold War periods in many unanticipated ways. The wartime[1] operational pressures revealed gaps in the Army's capabilities, and spurred an urgent drive from both the Army and the Department of Defense (DoD) to fill those gaps with new technology solutions.

What followed was a period of organizational creativity within the Army, where decisionmakers responding to the urgent operational needs from the field were also equipped in an unprecedented manner with a source of immediate flexible funding to respond to those needs: congressionally allocated supplemental funding. Perceiving both urgent needs and having in hand the resources to address them, the Army did not rely on the full formal structures of the Defense Acquisition System reflected in DoD Instruction 5000.02 on Operation of the Defense Acquisition System, because following that process would have taken too long to deliver the needed items.[2]

Instead, the Army and DoD developed, viewed from the highest level, two types of methods to perform rapid acquisition during this period:

- Establishing named, formally designated, rapid acquisition structures (i.e., processes and organizations)
- Applying the traditional acquisition structures in an unusual, non-"program of record," ad hoc manner.[3]

The Army used rapid acquisition methods to acquire a wide variety of capabilities, including weapon systems, vehicles, and individual equipment. In this work, our particular focus is on command and control (C2) systems, which are a subset of the larger

---

[1] By "wartime," we refer to an environment where the military is engaged to a significant and extended degree in combat (vice minor policing actions or small-scale special operations).

[2] *Army Strong: Equipped, Trained and Ready: Final Report of the 2010 Army Acquisition Review*, January 2011, states that it takes the acquisition system four years to go through the expected steps without producing anything at all.

[3] Programs of record follow DoDI 5000.02 and adhere to a standard set of decision processes and milestones. According to the Defense Acquisition University, a program of record is "recorded in the current Future Year's Defense Program (FYDP) or as updated from the last FYDP by approved program documentation (e.g., [APB, acquisition strategy, SAR])." Defense Acquisition University, "Glossary of Defense Acquisition Acronyms and Terms," Fifteenth Edition, December 2012. As of January 3, 2013: https://dap.dau.mil/glossary/pages/2492.aspx

category of information technology (IT)-based systems. Defense acquisition of IT systems has a number of unique attributes, including a high degree of reliance on commercial off the shelf (COTS) technologies that refresh in the commercial sector at a rapid rate, posing a significant risk of obsolescence at first fielding if delivery of a new IT system takes too long. C2 systems, as IT systems, also depend on other systems, and acquisition of C2 needs to consider interoperability. Yet at the same time, compared to other IT, a distinct aspect of the acquisition of C2 systems is that they are likely to be pulled by the urgent needs of war.

As shown in Figure S.1, the Army's reliance on these alternative mechanisms to support rapid acquisition has caused a new set of problems. These problems have included poor integration with existing systems, incurring additional operational risks, increased security risks, and a lack of sufficient support for the rapidly acquired systems. The question we were asked is, given the Army's recent experiences with rapid acquisition: How can DoD, and the Army more particularly, better perform rapid acquisition within the current policies and procedures?

The objective of our study was to discern how the DoD acquisition system can more rapidly develop, procure, and field effective C2 systems, and to provide the Army with recommendations to improve future rapid acquisitions of C2. In addition, we probed issues and challenges posed by rapid acquisition of C2. Finally, we identified factors that have enabled previous successful rapid acquisitions of C2.

Figure S.1. Study Motivation

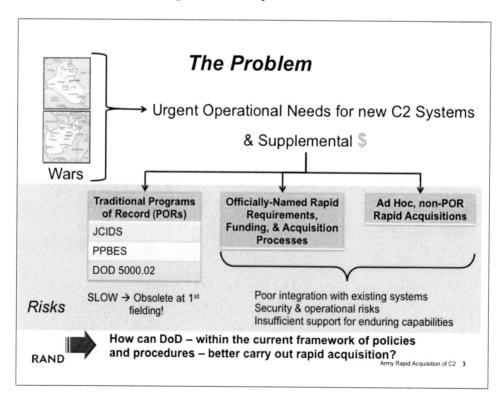

Our approach was twofold. First, we selected three rapid acquisitions of Army C2 systems to serve as case studies from which to derive lessons learned. Second, largely to enhance the analysis and conclusions from the case studies, we surveyed established rapid requirements, funding, acquisition, and transition processes. This secondary survey provided context, and it confirmed some of our case study findings and broadened our recommendations.

During our selection process, we considered as case study material 289 different programs suggested by the sponsor and other subject matter experts. From among that number we selected our three case study programs qualitatively on the basis of five main criteria:

- We could gain access to suitable information on the program to support the research task.
- The acquisition took place more rapidly than would normally occur, in the opinion of subject matter experts.
- The program would be considered a "success," in the sense that it was fielded and users liked it.
- The acquisition of the C2 program relied on nontraditional processes
- The set of programs represented a variety of program sizes.

After considering the assorted programs, and taking into account the fact that a particular air defense program called the "Fire Coordination Cell" motivated our sponsors to ask the original question behind this study, we selected the following three programs to serve in the case study approach:

- Joint Network Node (JNN): a satellite-based beyond-line-of-sight communication system
- Command Post of the Future (CPOF): a real-time decision support system
- Fire Coordination Cell (FCC): an air and missile defense system used to coordinate targeting

Figure S.2. Factors Enabling Rapid Acquisition of Army C2

Factors Enabling Rapid Acquisition of Army C2

		FCC	CPOF	JNN
Wartime-driven attributes	Answers need the Army perceives as valid and urgent		X	X
	Able to rely on a source of immediate flexible funding		X	X
	Built on matured technology (COTS/GOTS)	X	X	X
	Users accept less than 100% desired performance	X	X	X
	Army/DoD accept higher operational risk	X	X	X
Normal but affected by wartime	Leverages existing programs and documentation		X	X
	Demonstrates useful capability quickly	X	X	X
	Endorsed and advocated by operational champion	X	X	X
Common to successful acquisition programs	Coevolves the technology and the operational concept		X	
	Plans carefully a transition to normal acquisition status to ensure long term existence and support		X	X
	Includes ongoing user feedback during development	X	X	X
	Ensures personnel on program staff consistent			X
	Ensures contractor personnel consistent	X	X	X
	Retains senior retired military consultants as advisors		X	

RAND

Army Rapid Acquisition of C2   18

As a result of our analysis of these programs, we have identified a number of factors that enabled them to deliver a successful rapid acquisition of a C2 system for the Army. Figure S.2 lists those factors, grouped by row into categories. For each factor, the table indicates whether or not it applied to each case study program. In general, CPOF and JNN, which have transitioned into the long-term inventory of the Army, incorporated more of those factors into their acquisition processes than FCC, which had not developed a viable sustainment arrangement as of the original writing of this report (2009–2010).

Two of the three case study programs were responding to urgent needs from the theater, and for them, war was a driving factor. The top group of program attributes we judged to be directly or indirectly war-driven. The middle group were less affected by the war, but still influenced by it. The bottom group we would consider to be common across successful acquisitions, rapid or otherwise.

**Figure S.3. War Drives Many Factors Enabling Rapid Acquisition,
and the Factors Influence Each Other**

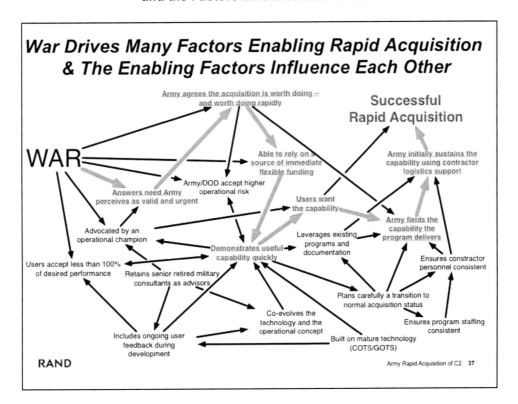

Figure S.3 takes these same enabling factors and shows their complex interactions. In effect, it can serve as a kind of map for future program managers seeking to conduct rapid acquisition. Green portions indicate the most essential flow of accomplishments for successful rapid acquisition. The figure shows how the enabling factors for rapid acquisition influence each other to create an institutional posture that is conducive to rapid acquisition success. Certainly, upon some study of this figure, it is possible to see that war is a major driver for successful rapid acquisition, influencing institutional motivations in a variety of ways. The urgency of a wartime environment enables traditional acquisition processes to be bypassed. War is a driver of many of the factors enabling rapid acquisition, and those factors depend upon and influence each other.

Figure S.4. Summary of Key Findings on Army Rapid Acquisition of C2

Summary of Key Findings

High-level champions are critical to the success of rapid acquisitions

Organizational and user flexibility enables rapid acquisition of new capabilities

- Wartime environments motivate bureaucratic flexibility
- For needed equipment, with good program office communication, users will tolerate operational risk and less than 100% of desired performance
- Ongoing user feedback enables iterative capability enhancement
- Rapid acquisitions require an immediate flexibly-tasked source of funding

A large variety of named rapid acquisition processes have supplied a number of wartime capabilities for DoD and the Army

- These designated processes have not met every rapid acquisition need of the Army
- The case study programs did not rely heavily on these named processes.

Relying on existing technology and documentation speeds acquisitions

- "Faster" can mean leveraging existing requirements, contracts, and documentation from other programs
- Rapid acquisitions require mature technology

C2 rapid acquisitions have ensured field support and sustainment via contractor warranties and eventual transition to an Army POR

- C2 rapid acquisitions have relied on multi-year warranties
- Transition planning and staffing consistency during the transition is essential

RAND   Army Rapid Acquisition of C2   **25**

For this study we also conducted a broad review of Army and Joint urgent need, rapid acquisition, and transition processes. From the case studies and informed by the review of existing acquisition processes, we identified a set of key findings (shown in Figure S.4) regarding how the Army has supported rapid acquisition of C2 during the last decade.

Our case study findings show that successful rapid acquisitions have relied on the patronage and support of highly placed individuals within the Army. These individuals substantiated the utility of a new capability as perceived by the rest of the Army organization, and supplied the required lobbying power to secure funding, support for development, and—especially—fielding.

In addition, the case studies have illustrated the types of flexibility required from the Army and DoD to support rapid acquisition of C2. The acquisition bureaucracy must be flexible in terms of business processes, and users must be flexible in terms of cost/schedule/performance tradeoffs.

The existence of war has been essential to successful rapid acquisition. War convinced the institutional Army that there was an urgent need for a new capability. War infused the bureaucracy with motivation for tolerating unusual process flexibility. War motivated Congress to supply the Army with a source of flexibly-taskable funding. Again, war motivated users to accept less than one hundred percent of the capabilities they had requested, and the Army to accept increased operational risks. Finally, war motivated operational champions to care enough about proposed solutions to problems to advocate for programs.

Another significant finding was that relying on existing technology and documentation sped the acquisition of C2 programs. The rapid acquisition program must build on mature technology. It can evolve the concept in constant feedback with users to increase the chances of delivering a useful capability. More surprisingly, programs can use contracts, requirements documents, and sustainment structures of existing programs to choreograph rapid initiation of a concept and ensure its fielding and sustainment.

DoD is currently planning to institutionalize rapid acquisition, that is, to plan its rapid processes so they persist in a structured and predictable way in the absence of large-scale conflicts.[4] As shown in Figure S.5, institutionalizing rapid acquisition poses both challenges and opportunities for the Army.

Challenges may impede DoD's push for institutionalization. For instance, it will be difficult to motivate (or justify) rapid acquisition in the absence of war. It will also be a challenge in the face of expected congressional skepticism to establish a flexible stream of funding to support rapid acquisition. Stakeholders for various traditional acquisition concerns will also push to have their interests more fully integrated, potentially slowing acquisitions.

---

[4] Also cited in the main body of the report, a June 14, 2012 DepSecDef memo on "Rapidly Fulfilling Combatant Commander Urgent Operational Needs" directs the department to establish policy and procedures to conduct rapid acquisition.

Figure S.5. Institutionalizing Rapid Acquisition Poses Both Challenges and Opportunities

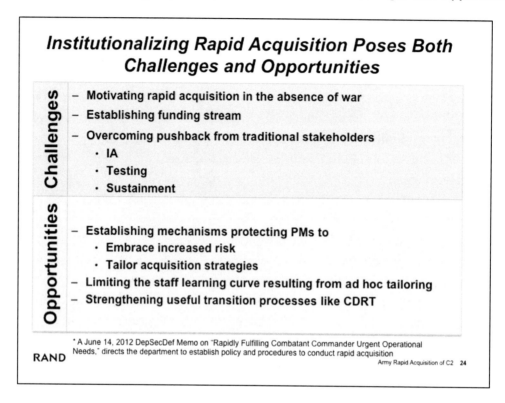

For the Army, however, institutionalization also provides some opportunities. It can establish mechanisms to protect program managers when they embrace increased risks within their programs, and help them tailor their acquisition strategies. Institutionalization also can minimize the required learning curves for staff seeking to replicate rapid acquisition techniques. It can, moreover, enable the Army to strengthen important processes to transition rapid acquisitions into the standard system as the program matures.

Figure S.6 contains our recommendations for how the Army can better conduct rapid acquisitions of C2 in the future.

The Army should regularly and systematically capture "Lessons Learned" from rapid acquisition.

One of our most easily actionable recommendations is that the Army should document its recent experiences in rapid acquisition to capture lessons learned and best practices, and develop metrics for program managers while the difficult-to-replenish reservoir of talent experienced in rapid acquisition expertise is still accessible and remembers much of what it has accomplished.

Figure S.6. Recommendations for the Army to Improve Rapid Acquisition of C2

> ## *Summary of Recommendations*
>
> - Regularly & systematically capture "Lessons Learned" from rapid acquisitions
> - Establish and support flexible mechanisms for funding acquisition efforts outside of the 2-yr. POM cycle, without supplemental funding.
> - Promote awareness of and strengthen existing Army processes for transition of non-PORs to official standing, such as CDRT
> - Establish ways to expedite testing to support rapid acquisition
> - Train the institution to expect PMs to tailor their acquisition strategy and mechanisms used
> - Enable PMs to prioritize and make trade-offs in meeting requirements based on user feedback
> - View acquisition documents, staff, and contracts of existing programs as potential enablers of rapid acquisition
> - Require PMs to assess planned rapid acquisitions for inclusion of the enabling factors discussed here
>
> RAND   Army Rapid Acquisition of C2   **26**

The Army should make it a priority to convince Congress to allocate flexible funding for rapid acquisitions of a needed type of capability on an ongoing basis.

The Army should seek ways to convince Congress to allocate some flexible funds to support rapid acquisition on a ready basis, even after current conflicts end. Funding for rapid acquisitions should be by type of activity, rather than "by thing." Because Congress prefers to retain its oversight and decisionmaking authorities, this is a potentially difficult case for the Army to make; it may be helped if the Army collects comparative data on the time necessary to initiate work on programs after an idea is conceived.

The Army should promote awareness of and strengthen existing Army processes for transition of non-PORs to official standing, such as CDRT.

The strengthening of these mechanisms should include establishing institutionally recognized mechanisms to *implement* the transition to a program of record (POR), not just establishing a mechanism to decide to do the transition.

The Army should explore ways to expedite testing in support of rapid acquisition.

How to tailor testing for a rapid program still seems an organizationally unsolved problem.

The Army should train the institution to expect program managers to tailor their acquisition strategy and mechanisms used.

Currently, program managers seeking to conduct rapid acquisition face considerable institutional resistance when planning to abridge or omit any of the standard parts of the 5000.02 acquisition process, as performing every aspect of the 5000.02 to full specification would slow the acquisition beyond the urgent needs of the Army.

The Army should make the requirements change process for rapid acquisitions easy and unencumbered to enable program managers to prioritize and make tradeoffs.

It should train its senior personnel, those developing and managing requirements, and program managers to empower program managers to prioritize requirements and make tradeoffs among them. The Army should also support program managers who significantly tailor their acquisition strategies to expedite acquisitions.

Both the institutional Army and program managers should view existing related programs as structures that can help acquire related capabilities quickly.

Existing programs have many of the sorts of documents and agreements that a new related program needs. In many cases, prior successful rapid acquisitions have appropriated those documented requirements, contracts, and support mechanisms, and also internal institutional Army documents, and modified them to include the new desired capability. The case studies in this research illustrate how the process of modifying these documents and agreements is less cumbersome than creating them anew.

In the course of future rapid acquisitions, we recommend that the Army require program managers to assess the rapid acquisition enabling factors from this report (shown in Figures S.2 and S.3), and to account for whether they have included those factors in their program appropriately.

In conclusion, the Army has considerable opportunities to improve rapid acquisition of C2 systems within the existing framework of policies and processes. By capturing lessons learned, implementing the recommendations in this document, and explaining to program staff the importance and interdependence of the factors shown in the guidemap for rapid acquisition of C2 shown earlier in Figure S.3, future rapid acquisitions may find it easier to replicate the success of past endeavors.

The Army has considerable opportunity to conduct and improve rapid acquisition of C2 and other systems within the framework of existing policies and processes. However, maintaining the wartime cultural and policy environment that enables and supports rapid acquisition in the absence of war is a significant challenge. We believe that the single most important action that the Army can take to institutionalize and improve upon existing rapid acquisition capability is to carefully capture the rapid acquisition experience of the last decade by fully documenting the program management and acquisition strategies of both successful and less so rapidly acquired systems, and make that documentation available to future program managers. A core lesson is understanding how the factors affecting success interact with and reinforce each other, and applying that understanding to the design of future rapid acquisition policies, processes, and programs.

(This page is intentionally blank.)

Acknowledgments

Special thanks to those in the Army's Project Manager Battle Command and Product Manager Tactical Battle Command offices, DARPA, and General Dynamics C4 Systems for providing us with pertinent background information, successes, and lessons learned for the CPOF program. In addition, we would like to thank those in Assistant Secretary of the Army for Acquisition, Logistics and Technology (ASA(ALT)) who provided us with initial contact information for CPOF.

The study authors thank the staff in both the Warfighter Information Network– Tactical (WIN-T) and the WIN-T Increment 1 offices for extensive helpful discussions, feedback, and provision of documentation in support of this study. We also thank LTC Douglas Smalls of ASA(ALT) for placing the authors in contact with program office staff at the beginning of the study.

The authors would also like to thank John Broussard and John Robinson in the Space and Missile Defense Battle Lab for assisting us in gathering information on FCC.

In addition, the authors thank the sponsors of this study, the Deputy Chief of Staff for Logistics (G-4), Director, Acquisition and System Management, ASA(ALT), and Director, Space and Missile Defense Battle Laboratory, Army Strategic Command, for their support.

Finally, the authors would like to thank the reviewers whose comments and suggestions greatly improved the final document: Hon. Claude Bolton, former ASA(ALT), and our RAND colleagues Jessie Riposo and Mark Lorell.

Any errors are the responsibility of the authors.

(This page is intentionally blank)

Glossary

1CD	1st Calvary Division (Army)
3ID	3rd Infantry Division (Army)
AMD	Air and Missile Defense Battalion
AR2B	Army Requirements and Resourcing Board
ASA(ALT)	Assistant Secretary of the Army for Acquisition, Logistics and Technology
ASD(NII)	Assistant Secretary of Defense for Networks and Information Integration
BFN	Bridge to Future Networks
BL	Battle Lab
BLOS	Beyond Line-of-Sight
C2	Command and Control
C3T	Command, Control and Communications–Tactical
C4	Command, Control, Communication, and Computing
CAIG	Cost Analysis Improvement Group
CDRT	Capabilities Development for Rapid Transition
CMU	Carnegie Mellon University
COCOM	Combatant Commander
COTS	Commercial-off-the-Shelf
CPD	Capability Production Document
CPOF	Command Post of the Future
DAB	Defense Acquisition Board
DARPA	Defense Advanced Research Projects Agency
DoD	Department of Defense
DOT&E	Director of Operational Test and Evaluation
DSB	Defense Science Board
DTIC	Defense Technical Information Center

FCC	Fire Coordination Cell
FY	Fiscal Year
GAO	Government Accountability Office
GOTS	Government-off-the-Shelf
GWOT	Global War on Terrorism
IBCS	Integrated Battle Command System
IED	Improvised Explosive Device
IOC	Initial Operational Capability
IOT&E	Initial Operational Testing and Evaluation
IT	Information Technology
IWN	Immediate Warfighter Need
JIEDDO	Joint Improvised Explosive Device Defeat Organization
JNN	Joint Network Node
JRAC	Joint Rapid Acquisition Cell
JUON	Joint Urgent Operational Need
LRIP	Low Rate Initial Production
MCS	Maneuver Control System
MDAP	Major Defense Acquisition Program
MRAP	Mine Resistant Ambush Protected (vehicle)
MSE	Mobile Subscriber Equipment
Non-POR	Non-Program of Record
OCO	Overseas Contingency Operation
OEF	Operation Enduring Freedom
OIF	Operation Iraqi Freedom
ONS	Operational Needs Statement
OSD	Office of the Secretary of Defense
PA&E	Program Analysis and Evaluation
PEO	Program Executive Office
PM	Program Manager

POM	Program Objective Memorandum
POR	Program of Record
PPBES	Planning, Programming, Budgeting, and Execution System
QRF	Quick Reaction Force
RDT&E	Research, Development, Test, and Evaluation
REF	Rapid Equipping Force
RFI	Rapid Fielding Initiative
RRF	Rapid Reaction Fund
RRTO	Rapid Reaction Technology Office
SMDC	Space and Missile Defense Command
TRADOC	U.S. Army Training and Doctrine Command
UON	Urgent Operational Need
USD(AT&L)	Under Secretary of Defense for Acquisition, Technology, and Logistics
USD(I)	Under Secretary of Defense for Intelligence
USD(P)	Under Secretary of Defense for Policy
USD(P&R)	Under Secretary of Defense for Personnel and Readiness
USMC	United States Marine Corps
WIN-T	Warfighter Information Network–Tactical

(This page is intentionally blank.)

1. Introduction

Figure 1.1.Rapid Acquisition of Army C2 Systems[1]

Rapid Acquisition of Army C2 Systems

Final

In this report we discuss the successful methods the Army and DARPA have used to rapidly acquire for the Army command and control (C2) systems in support of overseas contingency operations (OEF, OIF). We have prepared this report to assist the Army in planning and conducting future rapid acquisitions.

---

[1] The version of the briefing slides we use within this document differs somewhat from our original set of slides, to improve clarity and to address comments received in review.

Figure 1.2. The Problem

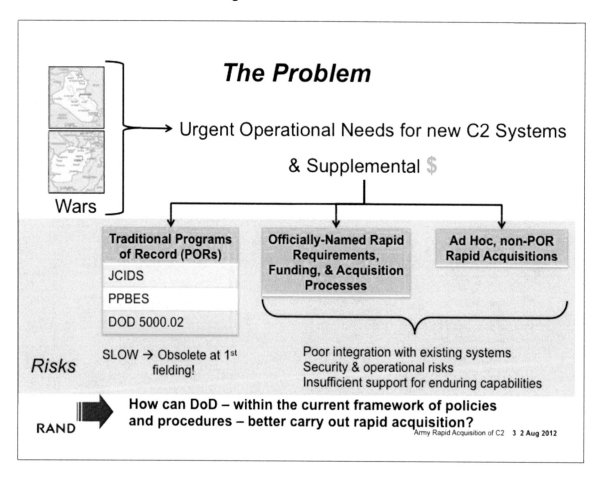

The Problem

Wars → Urgent Operational Needs for new C2 Systems & Supplemental $

Traditional Programs of Record (PORs)	Officially-Named Rapid Requirements, Funding, & Acquisition Processes	Ad Hoc, non-POR Rapid Acquisitions
JCIDS		
PPBES		
DOD 5000.02		

Risks

SLOW → Obsolete at 1st fielding!

Poor integration with existing systems
Security & operational risks
Insufficient support for enduring capabilities

RAND ➡ **How can DoD – within the current framework of policies and procedures – better carry out rapid acquisition?**

Army Rapid Acquisition of C2   3   2 Aug 2012

The conflicts in Afghanistan and Iraq drove the Army's recognition of a number of urgent operational needs. They also made available an unusual amount of flexible funding in the form of supplemental appropriations. If the Army had relied on the Department of Defense's normal acquisition processes, including the Joint Capabilities Integration and Development System (JCIDS) to establish formal requirements, the Planning, Programming, Budgeting, and Execution System (PPBES), and the DoD Instruction 5000.02, it would have been reasonable to expect the acquisitions to take fifteen years or more.[2] DoD realizes that this standard process is too slow to meet urgent operational needs.[3] Also, in the case of IT-based systems, because of rapid changes in

---

[2] Business Executives for National Security, "Getting to Best: Reforming the Defense Acquisition Enterprise," Defense Acquisition Archives, July 2009. As of July 2013: http://www.bens.org/document.doc?id=44

[3] See, for instance, the first paragraph of Jacques S. Gansler, "Final Report of the Defense Science Board Task Force on Fulfillment of Urgent Operational Needs," Memorandum to the Chairman, Defense Science Board, June 23, 2009.

computer technology, the traditional acquisition system risks delivering an obsolete capability even at first fielding. Aware of the need to expedite certain acquisitions, DoD has established a number of formal "rapid" acquisition enabling structures, including named processes and organizations, and we discuss and consider a number of these within this report. However, we note that the primary focus of this work has been the analysis of three programs we selected as case studies in rapid acquisition. In large part the programs we discuss in this report did not rely on formal rapid acquisition structures. Instead they performed rapid acquisition by using, in nontraditional ways, existing structures not specifically tied to rapid acquisition.

As a result, while the Army successfully developed and fielded the capabilities, it faced challenges associated with managing these processes and the C2 systems acquired through them. These challenges have included poor or incomplete integration with other systems, fewer security precautions, increased operational risks, and less ease and robust arrangements for the sustainment of those capabilities that may endure within the Army inventory.

Consequently, the Army now has a large number of different kinds of systems that are intended or desired by some users to be in inventory for the longer term, including C2 systems acquired to satisfy specific needs in theater, that have not been subject to any of the lifecycle planning associated with systems. In this study we examine how DoD and the Army, within the current framework of policies and procedures, has performed and can better carry out rapid acquisition.

Figure 1.3. Rapid Acquisitions of Command and Control Capabilities Face Challenges

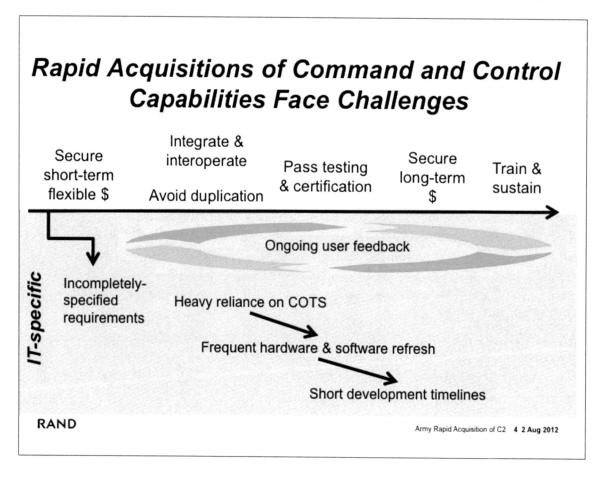

There are a number of common challenges for rapid acquisitions (as depicted above the left-to-right arrow in the figure above) regardless of the type of system to be acquired. To deliver a capability rapidly, systems must first secure flexible short-term funding to enable a start without delaying for two years to secure funding within the Army's normal Program Objective Memorandum (POM) development cycle. Then, development of the system must integrate the component enabling technologies, ensure interoperability with other systems as required, and avoid duplicating work from other efforts within DoD. The provisionally completed system must pass testing and certification requirements to be fielded, secure a source of long-term funding, train users, and provide for the future sustainment of the system. Regardless of the type of end application, all rapid acquisitions must surmount these obstacles.

In addition, command and control systems must also overcome obstacles that are particular to the rapid acquisition of IT-based systems, as shown in the grey portion of Figure 1.3. In many cases, C2 systems have incompletely specified requirements in the initial stages, and the nature of a system's capabilities and its intended uses evolves over the course of time with testing and the resulting user feedback. Moreover, rapidly acquired C2 systems have generally leveraged and integrated commercial off-the-shelf

4

(COTS) equipment, a practice that enables faster technology delivery but also entraps the military capability in the short development timelines necessitated by COTS frequent hardware and software refresh cycles.

Figure 1.4. Study Objective

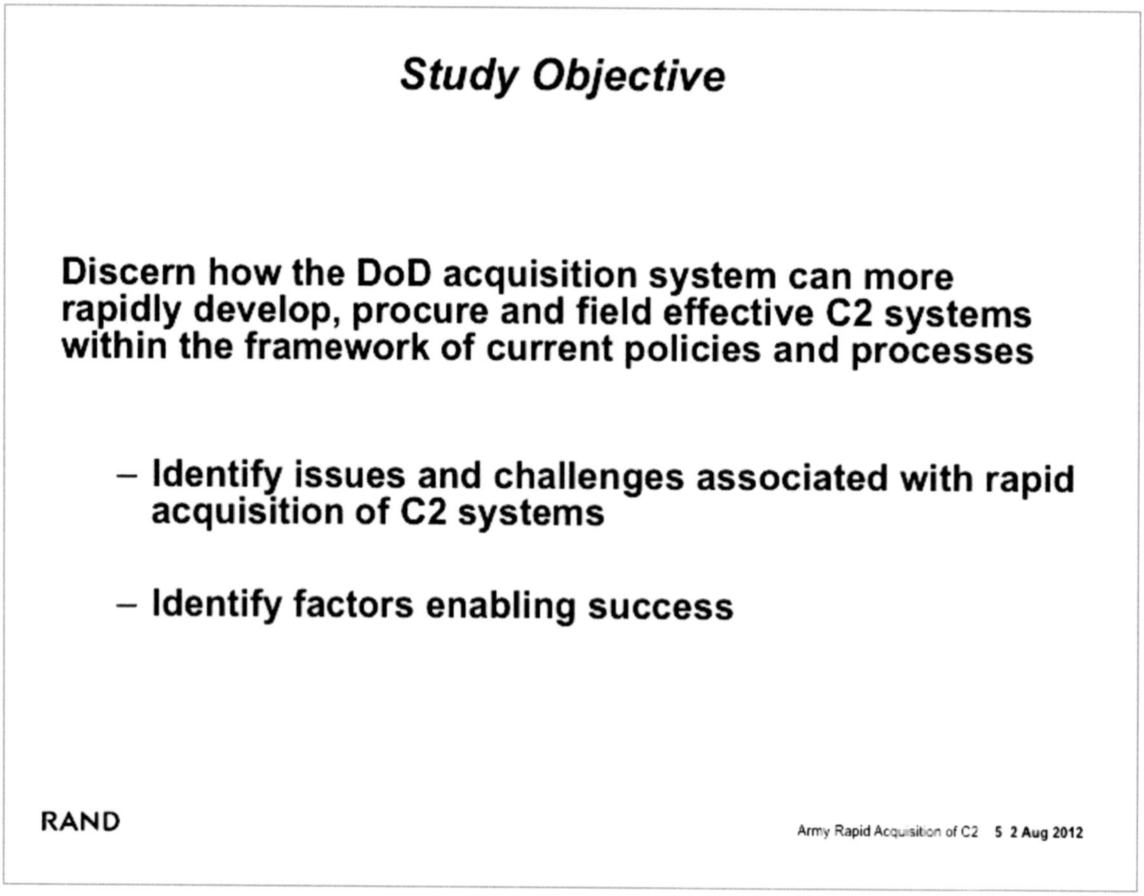

Study Objective

Discern how the DoD acquisition system can more rapidly develop, procure and field effective C2 systems within the framework of current policies and processes

- **Identify issues and challenges associated with rapid acquisition of C2 systems**

- **Identify factors enabling success**

RAND

Army Rapid Acquisition of C2   5 2 Aug 2012

Our research objectives were to analyze recent examples of rapid acquisitions to discern ways the Army can, within the current DoD acquisition system, more rapidly develop, procure, and field new C2 systems that deliver needed capabilities to the warfighter. Our goal was to identify issues, problems, and challenges associated with rapidly acquired C2 systems and rapid acquisition processes, and also lessons learned on factors enabling success. Ultimately, we want to provide insight into how rapid acquisition can be successfully accomplished in wartime as well as how to improve these rapid acquisition capabilities in the longer term.

Figure 1.5. Study Approach

Study Approach

- **Selected three C2 rapid acquisition programs to serve as case studies for lessons learned**
  - **On which info was accessible**
  - **Which relied on non-traditional processes**
  - **That garnered positive user experiences**
  - **Which represented various program sizes**

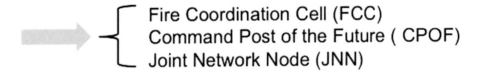

Fire Coordination Cell (FCC)
Command Post of the Future (CPOF)
Joint Network Node (JNN)

- **Surveyed established rapid requirements, funding, acquisition, and transition processes to inform analysis of the case studies**

RAND
Army Rapid Acquisition of C2   6   2 Aug 2012

Our approach for this study relied first on case study analysis of three C2 systems that were developed, procured, and fielded rapidly in response to urgent user requests. Second, we also surveyed formal established rapid acquisition processes within the DoD systems. In the course of our analysis, it was a revelation to find that the three systems we analyzed did not rely extensively on these established rapid acquisition processes to produce the delivered capability. Instead, the programs relied on ad-hoc application and tailoring of existing structures and mechanisms, but outside the processes that govern "programs-of-record" that follow the 5000.02, JCIDS, and PPBES processes and decision points.

To select the case studies we first solicited recommendations from a number of sources, including subject matter experts and the study sponsors. In all, we considered an array of 289 C2 systems as candidates for this study. One of the case studies we selected, the Fire Coordination Cell (FCC), was a choice of the study sponsor, as the problems this acquisition program encountered were the motivation for the study. From the 288 considered in addition, we chose two others:

- Command Post of the Future (CPOF)
- Joint Network Node Network (JNN-N)

We selected the set of three systems to serve as case studies based on qualitative consideration of the following criteria:

- We were positioned to secure the necessary information on the system to support the study and the analysis.
- The Army acquired the system via nontraditional requirements, budgetary, and acquisition processes.
- The system delivered a capability that a community of users liked and deemed useful.
- The system as a case study was such that it could illustrate the nature of contrasts and conflicts with the traditional acquisition system.
- The systems represented varying program sizes in terms of dollars invested and numbers of systems delivered.

Using the case studies, we sought to identify lessons learned for planning and managing future rapid acquisitions by the Army. A web-only companion document, RR-210-A (not available to the general public), discusses each case study in detail. This document relates the major points and lessons learned from the companion document. It also discusses the results of our survey of formal rapid acquisition institutional structures within DoD.

To conduct each case study we held interviews of many hours with a variety of senior program staff, including program managers, and civilian and military staff, over one or two days for each program. We also in some cases interviewed the contractor developers. While we would have preferred to include interviews with users as well, we could not do those interviews within the scope of this work. We selected people to interview based on their knowledge of the acquisitions under discussion. We attempted to interview representatives of major entities involved with the acquisition. For the interviews we had a very high-level set of topics for discussion, but they were not formally structured. In some cases, those interviewed had knowledge only of a particular aspect of the program, in which case the interview focused on the areas where they had knowledge.

In several cases those interviewed were also natural proponents, as the capability was something they had devoted much time and effort to developing and fielding. We did not interview representatives of the testing community, once again, subject to our resource constraints. We have attempted to balance our commentary by relying on reliable source documents and factual information wherever feasible.

In addition, to support the case studies, we requested available documentation from program offices to support understanding cost, schedule, and process dates and events. We discuss much of this detail in the companion report.

As we begin discussing successful rapid acquisition programs, we should first state how we define "successful" and "rapid." We have judged "success" in a qualitative manner based on evidence that the program fielded a useful capability. "Rapid" means that the time from initiation of development to initial fielding was relatively faster than a traditional acquisition program could achieve. By the measure of other experts, it is reasonable to expect such programs to take longer than a decade to field a useful capability. For instance, it has been stated:

> Defense acquisition revolves around 15-year programs, 5-year plans, 3-year management, 2-year Congresses, 18-month technologies, 1-year budgets, and thousands of pages of regulations.[4]

While we do address cost and schedule, we have not used metrics of cost growth and schedule slip in our definition of "successful" because, while those metrics are familiar to the acquisition community, they generally reflect the combined skill of the cost estimators, those estimating the initial schedule, and the program management team, contractor performance, and the influence of external factors. Wrongness in those factors does not necessarily impact the value of the system operationally. In standard acquisitions, that sort of error does much to impact the likelihood of a program cancellation or an underfunding decision, but in this work we did not see evidence that such problems had been a significant consideration for rapid acquisitions. Moreover, it was not clear that traditional cost or schedule estimates were done in the beginning of these programs, so it is impossible to assess whether any cost increase or schedule slippage occurred. In any case, at the beginning of two of the three systems, program managers did not anticipate how widely the systems would be fielded, so the total cost and schedule of the program would have been completely unknown. By fielding all three of the systems to everyone who needed them, the Army has cast its vote that the systems' capabilities and performance levels were worth at least as much as they cost. A different question, and one that we did not explore in our research, is whether the Army could have purchased each system for less. However, we would argue, were it true, that the value of the system to the Army is whether the Army has shown itself willing to use it and pay for it.

Rapid acquisition processes confront several challenging tradeoff issues. First, there is the obvious tradeoff between responding to an urgent operational need in a timely fashion compared with satisfying longer-term capability requirements. The response to a Urgent Operational Need (UON) will not always be useful as a longer-term capability. Second, there need to be tradeoffs available to program managers to reduce system capabilities in order to field the system more quickly. Similarly, rapid acquisition

---

[4] Business Executives for National Security, "Getting to Best: Reforming the Defense Acquisition Enterprise," Defense Acquisition Archives, July 2009.

inherently involves accepting some increased operational risk; there is not always time to test a new system fully prior to its fielding and operational use.

The rapidly acquired systems of interest in this research are non-programs of record (non-PORs), meaning that they are not part of the normal planning, programming, and budgeting process.[5] The processes used to acquire such systems are also nontraditional (or nonstandard). As a result, there is the issue of transitioning non-PORs to the mainstream.[6] The specific issues that arise here include technology transition, product transition, continued development, and sustainment and training. Interestingly, these issues are the same as those experienced by most systems or technologies as they move from the science and technology (S&T) base, through advanced technology development, prototyping, engineering and manufacturing development, and production and deployment. This set of issues is thus not unique to rapidly acquired non-PORs, but they may uniquely need to be "backfilled."

IT systems require different acquisition policies and processes, due to differences in their characteristics. Again, this is true regardless of whether the system is a rapidly acquired non-POR C2 system or a C2 system acquired through the normal acquisition process. The IT system characteristics that are different from traditional acquisition programs (i.e., aircraft, tank, or missile programs) include the inability to fully define requirements up front, the need for early and continuous user feedback to help define and refine requirements, and the fast turnover of both hardware and software elements of the system.

This suggests that there are lessons from the traditional acquisition process concerning technology and program transition and IT system acquisition that can inform the design and improve the performance of rapid acquisition processes. There may also be lessons from the Army's experience in rapid acquisition of C2 systems that can be used to improve IT system acquisition more generally.

---

[5] According to the Defense Acquisition University, a "program of record" is "recorded in the current Future Year's Defense Program (FYDP) or as updated from the last FYDP by approved program documentation (e.g., [APB, acquisition strategy, SAR]). If program documentation conflicts with latest FYDP, the FYDP takes priority." Defense Acquisition University, "Glossary of Defense Acquisition Acronyms and Terms," Fifteenth Edition, December 2012. As of January 3, 2013:
https://dap.dau.mil/glossary/pages/2492.aspx

[6] By "transition" in this report we refer to the transition of the program from using nonstandard acquisition processes to a state of conforming with the expectations laid out in DoDI 5000.02.

Figure 1.6. The Organization of This Report

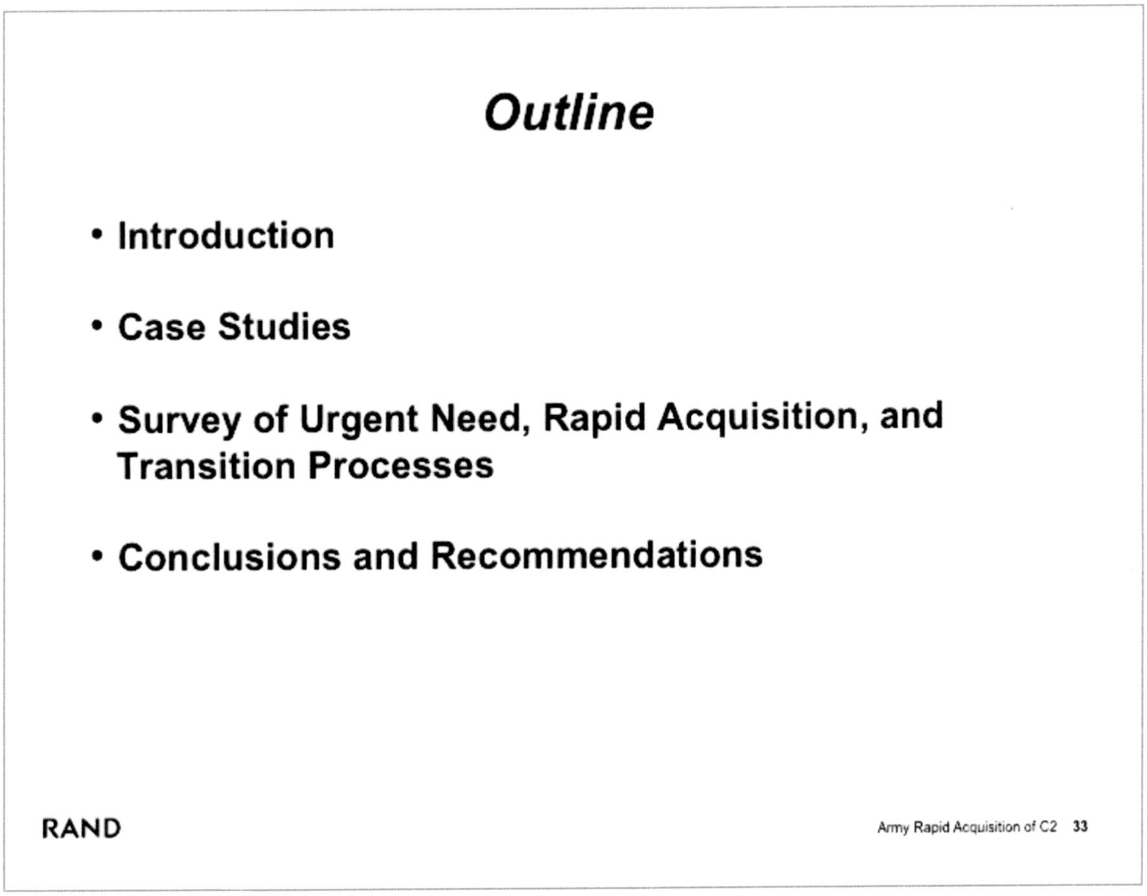

Outline

- **Introduction**

- **Case Studies**

- **Survey of Urgent Need, Rapid Acquisition, and Transition Processes**

- **Conclusions and Recommendations**

RAND   Army Rapid Acquisition of C2   **33**

The next chapter describes the results of our case study evaluations. These case studies are fully documented in the web-only companion report.

The third chapter provides an overview of existing DoD and Army urgent needs and rapid acquisition processes, as well as two Army processes that transition non-PORs into the mainstream. This review describes the policy environment and provides context for the case studies. It also identifies a set of factors affecting urgent needs and rapid acquisition processes, drawing on both our own analysis and several other assessments that were published in the same time period in which this research was conducted. In addition, we consider and discuss the case studies in the context of these established processes. Appendix A contains a description of each of the processes we examined and citations of supporting documents, as well as a longer discussion of the issues and challenges we observed.

Lastly, we conclude with our overall findings and recommendations for the Army on how to begin improving the management and oversight of non-POR programs and processes.

(This page is intentionally blank.)

2. Case Studies

Figure 2.1. The Section on Case Studies

Outline

- **Introduction**

- **Case Studies**

- **Survey of Urgent Need, Rapid Acquisition, and Transition Processes**

- **Conclusions and Recommendations**

RAND

Army Rapid Acquisition of C2   7

In this chapter we discuss the results of our analysis of case studies on three specific C2 systems the Army acquired rapidly within the past decade. We present the content of this analysis in much greater detail in the companion document. These systems all started as non-PORs and were "successful," in that they resulted in a fielded capability valued by the warfighter. In their different histories, we can identify certain program characteristics or factors affecting success that are common across the programs, and others that are unique to a particular program. These factors suggest ways the Army can enable a successful rapid acquisition in the future, as well as provide insight into how to improve existing processes.

The Three Case Studies Span a Range of Organizations, Funding Sources, Sizes, and Outcomes

	Fire Coordination Cell (FCC)	Command Post of the Future (CPOF)	Joint Network Node (JNN)
Owner/ Developer	Army SMDO Battle Lab	DARPA	Army (PEO C3T)
Focus	Unit reorganization	Urgent need	Urgent need
Type / Availability of Funding	limited funding from various sources	DARPA / supplemental / MCS funding	Supplemental funding / WIN-T
Program Size	> $3.5 million**	> $350 million*	~ $4.0 billion
Transition	No transition to POR but still deployed	Transitioned to existing POR (MCS) by satisfying unfulfilled reqts.	Transitioned to existing POR (WIN-T)
Participant in designated Army transition process	NO	YES – CDRT Iteration 2 (POR)	NO

All three provided "80% solutions" integrating COTS and GOTS equipment

RAND

*Includes RDT&E FY98-FY09, Other Procurement FY05-FY09

**Battle Lab spent approx. $3.5million for first 6 systems

Army Rapid Acquisition of C2   8

We examined three cases of rapidly acquired C2 systems—FCC, CPOF, and JNN-N—that illustrate a range of program characteristics; each has a unique story. However, there are also commonalities: All three programs used some combination of COTS and government off-the-shelf (GOTS) equipment to meet the expressed need. All three systems ended up achieving a militarily useful increment of capability, but not necessarily the initial objective (desired capability). All programs fielded systems to operational users.

Of the three programs, FCC is the only case that did not formally transition to a POR in some form, either enduring or sustaining (using CDRT terminology explained later in Chapter 3). Interestingly, FCC is also the only program that was not initiated as an urgent need from the theater and that did not have supplemental funding. These differences largely explain why FCC did not transition to a POR.

The case studies on three programs—JNN, FCC, and CPOF—are written up in detail in a web-only companion report that recounts the motivations behind and the historical narratives associated with each program.[11] It also provides more supporting detail for each of the lessons learned, findings, and recommendations in the present report.

[11] See Shara Williams, Jeffrey A. Drezner, Megan McKernan, Douglas Shontz, and Jerry M. Sollinger, *Rapid Acquisition of Army C2 Systems: Case Studies*, Santa Monica, CA: RAND Corporation, RR-210-A, 2013. Not available to the general public.

Figure 2.3. Fire Coordination Cell

Fire Coordination Cell (FCC)

Collaborative environment for Air and Missile Defense (AMD) battalion operations that provides situational awareness for Avenger and PATRIOT batteries to coordinate targeting

- SMDC Battle Lab (BL) built FCC prototype in ~4 to 6 months
  - **Mix of COTS and GOTS hardware and software**
  - **Server, laptop computers, projector, and screens**

- 7 FCCs deployed as of 2010, plus 1 FCC for training at Ft. Bliss
  - **First set of equipment delivered to 1-44 AMD Battalion October 2005**
  - **Cost about $400k per system**
  - **BL spent ~$3.5M for first 6 systems**

Source: Robinson, FCC Brief for IAMD, 6 Mar 08

RAND

Army Rapid Acquisition of C2   **9**

Our first case study examined the Fire Coordination Cell (FCC), a system that provides a collaborative environment to enable Avenger and PATRIOT batteries to coordinate fire during an air and missile defense engagement. FCC was the motivating case for this research. The need for FCC arose from the creation of composite air defense battalions that merged units operating the Avenger and PATRIOT systems. The Avenger provides short-range air defense, and the PATRIOT provides long-range air defense capabilities. In the past, control of the two systems was performed by two separate sets of personnel in separate locations. The original concept behind FCC was to create a single integrated fire control cell for both the Avenger and PATRIOT systems. The final system as implemented by FCC facilitated only coordination rather than control. This compromise in capability level represented an example of the warfighter accepting a "good enough" or "80 percent"[12] solution in order to acquire a useful capability quickly.

---

[12] By "80 percent" we refer in a notional manner to a level of performance that is sizable and valuable to the user, but not all of the desired capability, either in the number of functions or the level of performance.

The warfighters were willing to accept this compromise because having a partial capability was much better than having nothing.

In a period of just four to six months, the Space and Missile Defense Battle Lab at Space and Missile Defense Command (SMDC) developed and fielded FCC at a relatively low cost (less than $4 million total, or $400,000 per system), as a mixture of COTS and GOTS hardware. The only "new" subsystem in FCC was the middleware connecting the operating system and the applications. The Battle Lab developed the middleware and managed the program. As of the writing of this report in 2010, seven FCC units had been procured since fall 2005, plus one additional system used for training, located at Fort Bliss.

Figure 2.4. FCC History

Fire Coordination Cell (FCC) History

- In 2005 air and missile defense (AMD) battalions needed improved situational awareness for C2
    - Combined Patriot and Avenger batteries
    - Created by MG Vane

- TRADOC invited contractors to propose a concept at the Roving Sands 2005 biennial exercise
    - Only SMDC Battle Lab (BL) proposed a concept
    - No contractors participated

- MG Vane
    - Felt FCC met most needs for the AMD battalions
    - Intended FCC as an interim solution

- Anticipated in 2016 is the much more costly existing POR -- Integrated Air and Missile Defense Battle Command System (IBCS)

- As of 2010 FCC was not a program of record and had no dedicated funding for sustainment
    - Transition attempts were unsuccessful
    - AMD battalions have paid for FCC O&M out of unit funds

RAND

Army Rapid Acquisition of C2 **10**

This chart summarizes the history of the acquisition of FCC. In 2005, (then) MG Vane stated the need for FCC. MG Vane, as Commanding General, U.S. Army Air Defense Center and Ft. Bliss, merged Avenger and PATRIOT air defense batteries into a single organizational structure—a composite air and missile defense (AMD) battalion. The composite battalions needed to operate both systems simultaneously, merging target tracking and fire control functions. U.S. Army Training and Doctrine Command (TRADOC) invited contractors to propose concepts for AMD battalions that could be demonstrated at the then upcoming Roving Sands 2005 exercise. Only the Army SMDC Battle Lab in Huntsville, Alabama responded; contractors did not respond. One hypothesis suggested for why contractors did not respond is that FCC was not a program of record with specified funding in the budget.

The SMDC Battle Lab, which did respond, quickly designed and built the prototype using a combination of COTS and GOTS hardware and software, and also developed the "middleware" that tied it all together. The system did not allow actual control of one system using the other's fire control, but rather enabled information from both to be displayed on a common screen, thus facilitating coordination. Since this setup met most of the new AMD battalion's needs,[13] the program was authorized to proceed. MG Vane intended FCC as an interim solution.

As of the writing of this report, FCC is not a program of record and has had no dedicated funding for development, procurement, and sustainment.[14] Funding was cobbled together annually by the TRADOC Capability Manager–Lower Tier (TCM-LT) personnel from various sources, including Army G-8 Force Development Air Defense, one or more program managers, and other end-of-year funds. Battle Lab attempts to transition the program to the Army's traditional acquisition community failed, predominantly because the program had no identified funding that would transfer along with responsibility for the program.

According to the Battle Lab, they developed FCC faster and more cheaply than the traditional acquisition process could have done.[15] The first prototype took less than six months to develop, and the first six FCC systems cost approximately $3.5 million. In comparison, the Army has paid two contractors approximately $12.5 million each to develop a concept for the Integrated Battle Command System (IBCS), expected in 2016. IBCS is intended to provide an integrated air defense capability similar in some ways to FCC, but with improved integration and expanded capabilities, including fire control.[16]. While FCC did not transition into the mainstream, it is deployed and continues to be used by the AMD battalions. Therefore, the Army must sustain FCC through at least 2014. As of the writing of this report, the AMD battalions have been sustaining FCC using their own unit funding.

---

[13] Some early program participants indicated that FCC met 90 percent of the need of the new AMD battalion. This is clearly a subjective estimate that cannot be independently verified. The important point is that the users perceived that FCC satisfied most of their needs, based on their use of the system in the Roving Sands 2005 exercise.

[14] We do not know why FCC was unable to receive funding from the Army.

[15] For instance, in *Army Strong: Equipped, Trained and Ready: Final Report of the 2010 Army Acquisition Review,* January 2011, it states that it takes the acquisition system four years to go through the expected steps without producing anything at all.

[16] As of 2012 it appears, based on publicly available information, that the command and control portion of IBCS is termed "FCC." Based on this work, which predates that information, we cannot say whether this term derives from the FCC technologies described here or is just a term that has been reused by IBCS to describe an independently developed capability. See AAR, "Integrated Technologies." As of December 29, 2012:
http://www.aarcorp.com/integrated_technologies/products/products.htm

Figure 2.5. FCC Lessons Learned

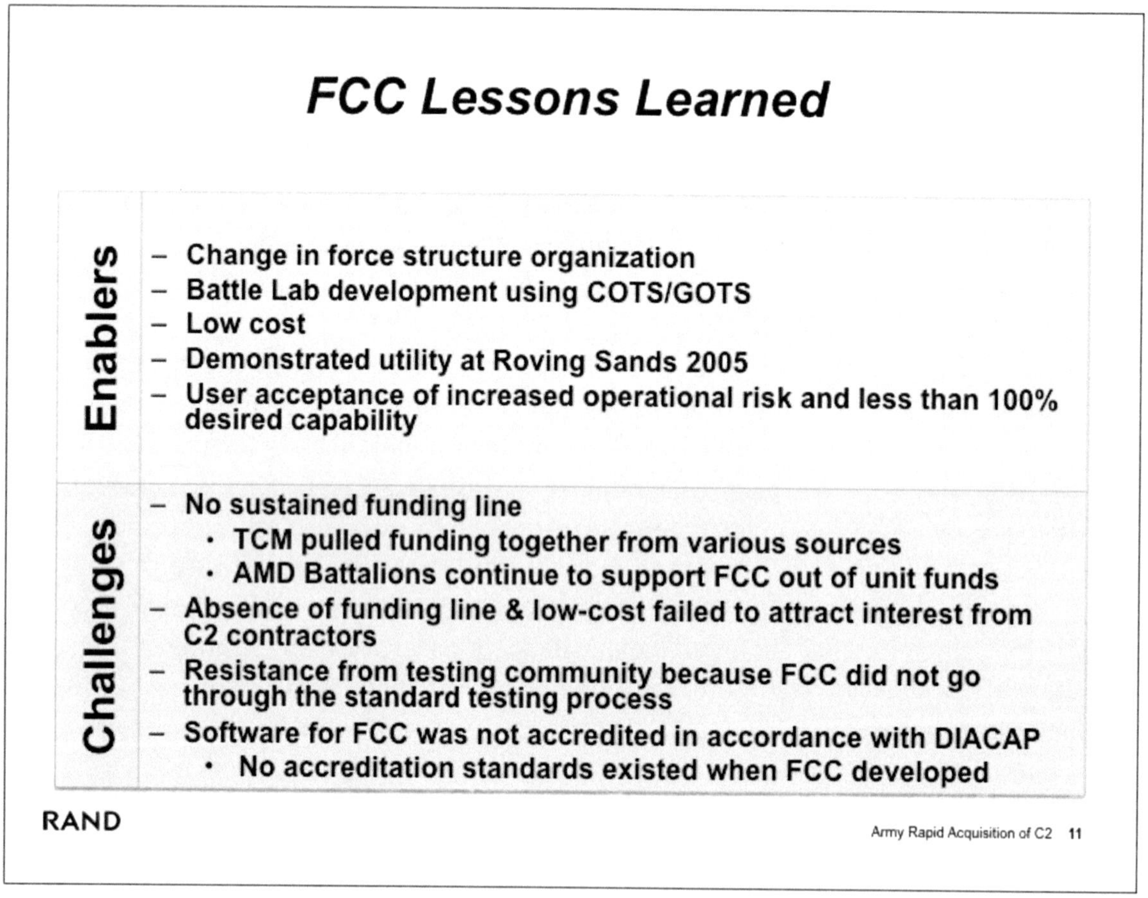

FCC Lessons Learned

Enablers
- Change in force structure organization
- Battle Lab development using COTS/GOTS
- Low cost
- Demonstrated utility at Roving Sands 2005
- User acceptance of increased operational risk and less than 100% desired capability

Challenges
- No sustained funding line
  - TCM pulled funding together from various sources
  - AMD Battalions continue to support FCC out of unit funds
- Absence of funding line & low-cost failed to attract interest from C2 contractors
- Resistance from testing community because FCC did not go through the standard testing process
- Software for FCC was not accredited in accordance with DIACAP
  - No accreditation standards existed when FCC developed

RAND

Army Rapid Acquisition of C2   11

As a rapid acquisition, FCC benefited from a number of enabling factors. First of all, the change in force structure organization to create integrated AMD battalions generated a new need and user base for the capability. As an IT-based system, FCC could rely on integration of COTS and GOTS capabilities to deliver the needed system at a low cost. Thirdly, the Battle Lab proved the system could work at the Roving Sands exercise in 2005, providing a venue for demonstrating the usefulness of the capability to the users.

However, FCC encountered a number of challenges, the chief of which was the lack of a dedicated funding line. The TRADOC Capabilities Manager (TCM) assembled funding for the capability from a number of sources. The lack of funding and institution support for the system failed to attract contractors to develop the capability. Moreover, FCC encountered resistance from the testing community because it did not go through the normal testing process. Because FCC was an IT system, some people in DoD also had the expectation that FCC should go through DoD Information Assurance Certification and Accreditation Process (DIACAP), but the standards had not been developed at the time FCC was developed. Lack of accreditation can introduce interoperability and information assurance risks, an example of the warfighter assuming greater operational risk than is done in a traditional acquisition program.

We can learn a number of lessons from the Army's experience with FCC. FCC demonstrated that if users are willing to accept performance compromises, by relying mostly on COTS equipment, C2 acquisition can occur successfully over a short development time period at low cost to satisfy new or different C2 requirements. For success, FCC relied on critical support from a high-level champion: MG Vane. One of the key factors affecting success of FCC was the use of the Roving Sands 2005 exercise to experiment with concepts and demonstrate both feasibility and usefulness. In contrast to the other cases we examined, the need for FCC was not generated by a UON, but rather by a force structure organizational change. The Roving Sands exercise played the same role that early operational use in theater plays in the other cases we examined. However, FCC's lack of validation by other parts of the Army organization left it with substantial challenges in securing funding and ensuring longer-term support. Sustaining systems developed like FCC requires formal decisions from senior leadership about how long systems will be fielded and whether to provide dedicated funding for the sustainment period. At the time of the writing of this report, FCC had not received the necessary support from the institutional Army.

Figure 2.6. Command Post of the Future (CPOF)

Command Post of the Future (CPOF)

CPOF is a real-time decision support system

- **Provides situational awareness**

- **Collaborative tools for tactical decision making, planning, rehearsal, and execution management**

- **CPOF rapidly processes and displays combat information from other supporting Army Battle Command Systems (ABCS).**

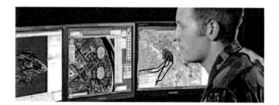

Sources:
Text : PM Battle Command
Picture: General Dynamics C4 Systems

RAND

Army Rapid Acquisition of C2   **12**

Command Post of the Future (CPOF) has become an important system in theater. It is a decision support system providing real-time situational awareness and collaboration among operators both within an echelon and among echelons at different command levels. CPOF processes and displays information from other Army C2 systems, and reduces transport risk in theater because commanders and subordinate units can share data and collaborate without having to travel.

Figure 2.7. CPOF History

CPOF History

	Late 1990s	DARPA developed CPOF concept and technology
	Sept 1999	Global InfoTek, Inc. awarded contract for DDT&E and demonstration support
"Rapid" period	**Jul 2003**	DARPA PM Ryan Paterson brought in to end CPOF; transition to Services was unavailable
	Aug-Sept 2003	Senior military consultants introduced CPOF to Army MG Peter Chiarelli (1CD), and he expressed strong interest
	Mar 2004	DARPA team deployed systems throughout Chiarelli's 1CD in Iraq, and critical operational need determined due to growing IED risks
	May 2004	Transition MoA between Army and DARPA signed
	Jan 2006	CPOF added as a "technology insertion" into Maneuver Control System (MCS) program
	June 2007	General Dynamics C4 Systems awarded contract for support and engineering services
	Jul 2009	Nearly 6,000 CPOF system clients fielded

RAND

Army Rapid Acquisition of C2 **13**

In comparison to FCC, the CPOF story is more complex. As shown above, the program originated at Defense Advanced Research Projects Agency (DARPA) in the mid-to-late 1990s. DARPA contracted with relatively small firms to foster innovation. In 2003, CPOF had reached the final stages of its lifecycle at DARPA without a clear transition path to a service. The final DARPA program manager was given direction either to transition the system to a military service or close down the program.

Senior military consultants, brought on board by DARPA in the role of "users" during early concept exploration and system development, introduced CPOF to then Army MG Peter Chiarelli of the First Cavalry Division (1CD), who expressed strong interest in the system after a demonstration. At the time, MG Chiarelli's 1CD was preparing for deployment to Iraq. A critical operational need for the collaboration capability was determined based on the growing threat of improvised explosive devices (IEDs).

A Memorandum of Agreement (MoA) between the Army and DARPA was signed in May 2004, concurrent with initial fielding in Iraq with the 1CD. Within two years, CPOF transitioned from DARPA to the Army as a "technology insertion" into the Maneuver Control System (MCS) program. The MCS program had unfulfilled requirements related to collaboration that CPOF satisfied, and introducing it as a technology insertion or modification allowed the program office to avoid creating the paperwork associated with a new program. As MCS had already passed Milestone C, it was not necessary for CPOF to prepare separate paperwork to support future milestone decisions.

At the time this research was conducted (June 2009–September 2010), CPOF had essentially taken over the MCS program largely because CPOF was perceived to be more user friendly and to have better operational capabilities. The demand for CPOF grew steadily in theater as its utility was demonstrated and its capabilities evolved. There were nearly 6,000 fielded clients of the CPOF system, and strong field support for training and sustainment.

Figure 2.8. CPOF Lessons Learned

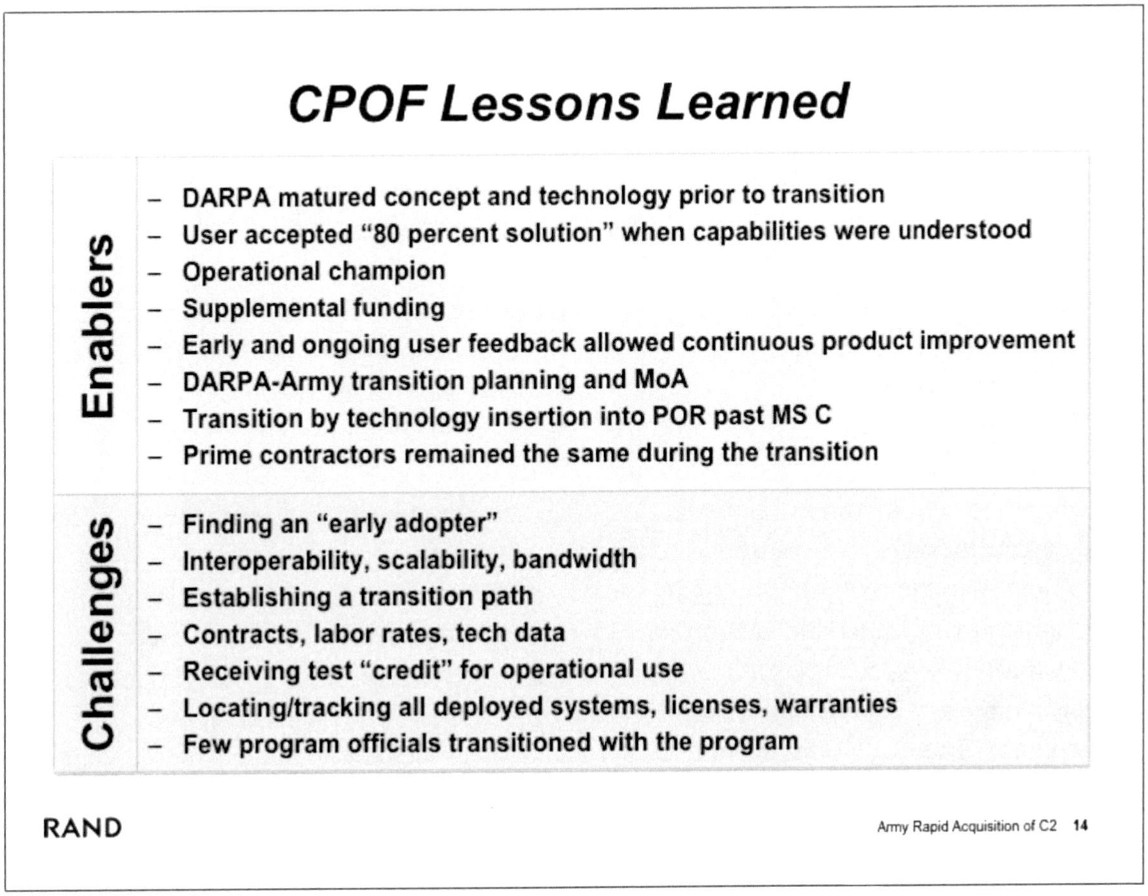

CPOF Lessons Learned

Enablers
- DARPA matured concept and technology prior to transition
- User accepted "80 percent solution" when capabilities were understood
- Operational champion
- Supplemental funding
- Early and ongoing user feedback allowed continuous product improvement
- DARPA-Army transition planning and MoA
- Transition by technology insertion into POR past MS C
- Prime contractors remained the same during the transition

Challenges
- Finding an "early adopter"
- Interoperability, scalability, bandwidth
- Establishing a transition path
- Contracts, labor rates, tech data
- Receiving test "credit" for operational use
- Locating/tracking all deployed systems, licenses, warranties
- Few program officials transitioned with the program

RAND

Army Rapid Acquisition of C2 14

The CPOF program faced a set of challenges that many DARPA programs face, including finding a transition path and an early adopter within a military service, and the rather mundane but important incompatibility between Army and DARPA policies on contracting, appropriate labor rates, purchase of technical data rights, and other administrative details. There were significant data rights (or intellectual property rights) associated with the use of COTS software. The Army was accustomed to paying lower labor rates, which made it difficult to transition contractors from DARPA to the Army. Additionally, no DARPA program officials transitioned to the Army program; the lack of program staff continuity caused some problems during transition. Initially, the technology insertion of CPOF into MCS engendered resistance. However, this problem was mitigated somewhat by the Army's efforts to transition to Army support the contractors who developed the system for DARPA.

Other significant challenges were associated with satisfying the wartime UON. Interoperability, scalability, and bandwidth issues presented significant unknowns as the system deployed throughout the Army because DARPA did not design the system to meet Army specifications in these areas. DARPA worked quickly to the help the Army resolve these critical issues during the transition period. Though tested extensively as part of its iterative development, CPOF still did not meet traditional testing requirements, and it appears that no "test credit" was given for its years in operational use when the formal test programs began after transition to the Army. In addition, DARPA did not have a formal system to track the location and configuration of each deployed CPOF system, along with their applicable licenses and warranties. There was also no "Army standard" logistics or training documentation for the Army to rely on. Both of these documentation issues proved to be challenges for the Army, since their fielding effort quickly expanded from small-scale to large-scale.

The experience of the CPOF program also demonstrated a set of factors that enabled the success of rapidly acquired C2 systems. Perhaps most importantly, DARPA matured the concepts and technologies, unconstrained by service IT acquisition rules. This included the use of senior retired military consultants in an iterative (or incremental) development approach from the beginning of the program at DARPA in both the development and fielding of the system. This helped to identify and refine requirements and capabilities and to give users a better understanding of how to use the system beneficially in the field, thus allowing the coevolution of technology and operational concepts. Senior leadership support was also critical, in this case in the person of then MG Chiarelli, who became an operational champion for the system; his experience using the system was an important factor in attracting additional users and Army-wide acceptance. The availability of funding through the supplemental budgets that did not compete with other budget priorities was also important.

DARPA and Army program personnel eventually made CPOF a technology insertion into the MCS, a creative use of existing requirements documents, contracts, and acquisition strategies rather than starting a new program. Perhaps most important, with proper communication from the program office, users were willing to accept a "good enough" "80 percent solution." The lesson here, especially applicable to IT-based systems, is that early and continuous user feedback can help identify and prioritize capabilities, which can then be added as the system matures.

CPOF benefited tremendously from institutional support and funding. A well-placed operational champion, a highly placed member of the military, advocated for the system and supported it in its vulnerable initial stages. In addition, it was able to rely on supplemental funding to provide a flexible source of funds.

Transition planning was critical to CPOF's success. DARPA and the Army signed a memorandum of agreement (MoA) to govern the transition, and DARPA ensured that the Army provided funding to support the program after its transition. In spite of obstacles with differing labor rates, the Army also aided the successful transition by ensuring that the contractor personnel remained the same during the transition, and the CPOF transition team involving both DARPA and the Army was aggressive and innovative.

The lessons learned from the CPOF program case study are that the Army can support a rapidly scaled acquisition when there is institutional recognition of the need and a flexible source of funding. That recognition enables the program to overcome obstacles and ensure fielding. Relying on COTS-based technologies aids fielding a capability in a rapid manner. The experiences of CPOF suggest that rapid acquisitions should seek operational champions to support them within the Army institution, especially in the early periods. C2 programs seeking to develop a useful capability quickly should consider using the technique CPOF used: to coevolve the technology and the operational concept with the participation of users, and to concentrate on meeting the subset of requirements most important to users.

Future rapid acquisition programs can also take a number of lessons from CPOF's transition to the Army. Programs can seek to transition by inserting themselves into other related programs. In particular, if a related program has already passed Milestone C, the technology insertion and subsequent acquisition may require considerably less of an oversight and paperwork burden.

As a final lesson, the CPOF case study highlights the importance of retaining a consistent, highly qualified, and experienced team of personnel to perform rapid acquisition successfully. The senior military advisors to the initial DARPA program, the DARPA program staff, the Army transition staff, and the contractors performing the acquisition all had the right experience and qualifications, and the Army ensured that the contractor staff remained consistent through the transition.

Figure 2.9. The Joint Network Node (JNN) System

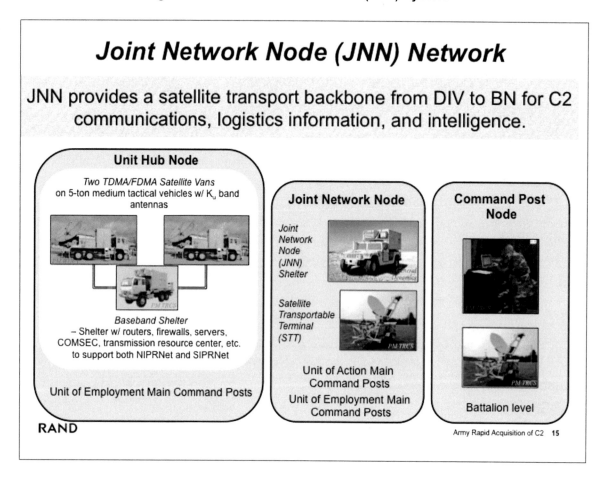

Like CPOF, JNN became an important system in theater. Given the demand for communication and data services, the program got very large very fast, relative to most systems rapidly acquired to meet an urgent need. JNN can be characterized as more of an infrastructure system than an end-user system. It was intended to provide a satellite-based, non-line-of-sight/over-the-horizon communications capability from Division to Battalion that featured set up at the "quick halt." The program relied on existing subsystems and components—COTS and GOTS—to provide that capability quickly.

For JNN, defining a "unit" (as in unit cost) was difficult; different military units receive different mixes of equipment based on the specific needs their mission in theater requires, and the equipment already in theater. The composition of JNN-N changed somewhat over time: single shelter switches, baseband nodes, and joint network nodes were all significant elements of the deployed system.

Developing concurrently with the acquisition of JNN, the Warfighter Information Network–Tactical (WIN-T) was a more ambitious, conceptually related program of record intended to eventually provide a mobile beyond-line-of-sight communication capability. However, WIN-T's schedule did not intend to field a capability until later in the decade, and WIN-T suffered delays in meeting its requirements.

Figure 2.10. JNN History

JNN History

Spr 2003	The march on Baghdad showed Army bandwidth and beyond-line-of-sight (BLOS) communications needs not supported
2004	PEO C3T, BCBL at Fort Gordon, and others delivered an initial COTS-based commercial K_u band satellite communications system "within weeks."
Jan 2005	3ID deployed to OIF with the first complete "JNN" network
Feb 2006	Deputy Chief of Staff G3/5/7 decided JNN should go to all units possibly deploying in '07-'09
Jun 2007	ADM merged JNN into WIN-T

Early pictures of JNN Diane Concepcion, *JNN Network Logistics*, PM TRCS, 2005

RAND

Army Rapid Acquisition of C2 **16**

Figure 2.10 summarizes the major events in the acquisition of JNN-N. The urgent operational need for a beyond-line-of-sight (BLOS) communication capability became apparent during the first weeks of the war in Iraq in 2003. Advancing Army units outpaced the capabilities of their existing communications equipment, MSE and Tri-TAC, which were not designed to support mobile, highly dispersed forces. In addition, the 3rd Infantry Division (3ID) realized a similar capability gap in its training exercises prior to deployment. An Operational Need Statement (ONS) was submitted and validated in 2004. Interestingly, a number of traditional organizations within the Army, including the Program Executive Office for Command, Control and Communications–Tactical (PEO C3T) and the Battle Command Battle Lab, collaborated to put together a system that would meet emerging needs. The Army redirected a contract from an existing program—Area Common User System Modernization—to enable some of the initial work, and then expanded it through multiple engineering change proposals. The validation of a Capability Production Document (CPD) from Bridge to Future Networks (BFN) was also used as authorization for JNN production. This is another example of the creative use of existing structures to help rapidly acquire a capability.

The first complete set of JNN equipment deployed to Iraq with the 3ID in early 2005. By early 2006, the Army decided JNN-N should field with all deployed units. Because JNN equipment was now going to essentially the entire Army, the size of the program, as measured in dollars spent, grew very high very rapidly.

An interesting controversy evolved over whether JNN was an ACAT 1 program, given that its funding quickly surpassed the dollar threshold defining an MDAP (major defense acquisition program). This issue had implications for production authorization without completion of testing and a "Beyond LRIP" report, as required by DoD 5000.02 for MDAPs. For JNN, the Army appears not to have asked for full rate approval or low rate initial production (LRIP); for that matter, for the initial period it omitted any formal decision-making milestone, but then briefed DOT&E (Director of Operational Test and Evaluation) asking for what it termed a "Limited Production Decision." The Army allocated funds and bought the equipment without going through any of the normal approvals and oversight. JNN used supplemental funding and was not a program of record, and so did not technically fall under DoD 5000.02 processes—according to some. Other DoD officials disagreed, however. Eventually, the "JNN law" was approved by Congress; it said if DoD allocates funds or proceeds to operational use without conducting IOT&E (Initial Operational Testing and Evaluation) and submitting a Beyond LRIP report, it must accomplish both "as soon as practicable."[17] The law closed the loophole of just skipping that acquisition phase entirely.

JNN eventually became Increment 1 of a restructured WIN-T program. At about the same time that controversy of whether JNN was an ACAT 1 program and subject to traditional oversight, the WIN-T program breached its cost thresholds. DoD and the Army simultaneously solved the problems of providing structure and oversight to the large JNN-N acquisition and the struggles of WIN-T to deliver a capability to users at an affordable cost in a timely manner by restructuring WIN-T and inserting JNN as its first increment of capability.

---

[17] John Warner National Defense Authorization Act for Fiscal Year 2007, P.L. 109-364.

Figure 2.11. JNN Lessons Learned

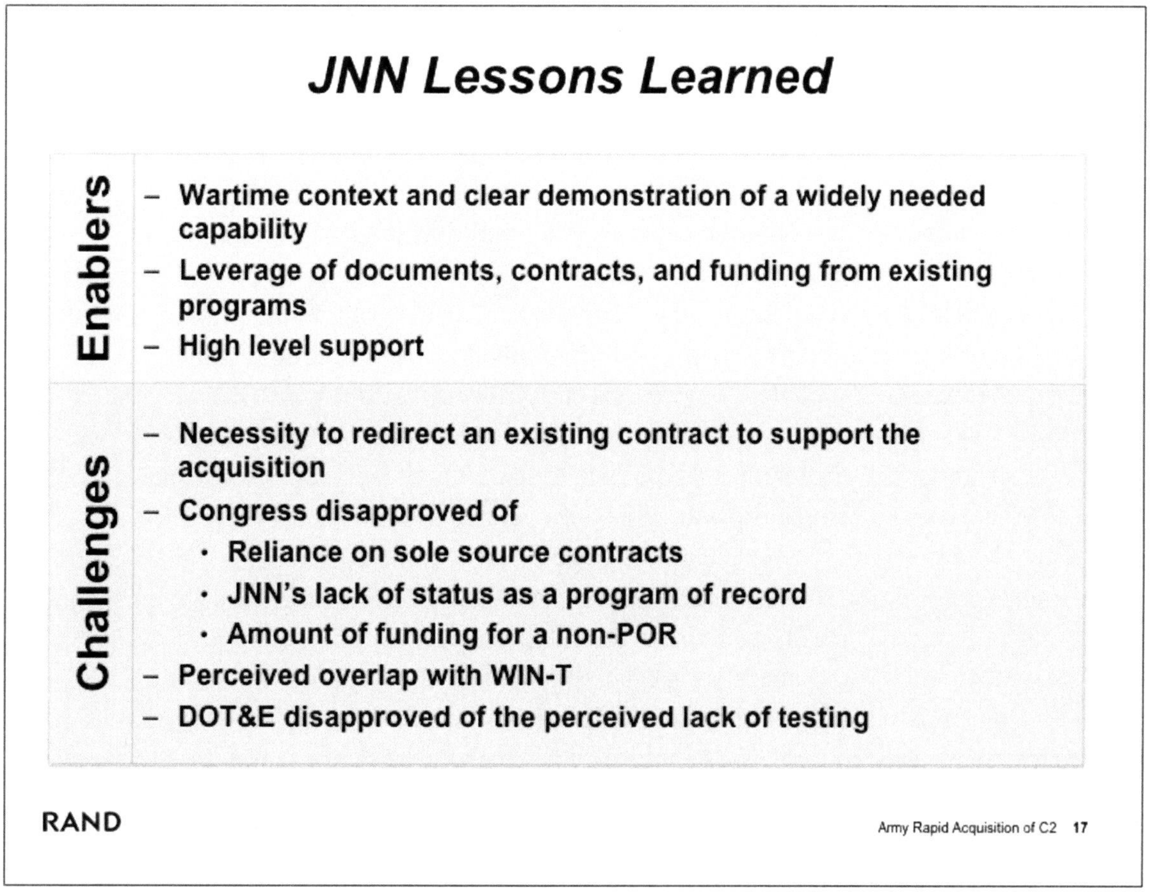

JNN Lessons Learned

Enablers
- Wartime context and clear demonstration of a widely needed capability
- Leverage of documents, contracts, and funding from existing programs
- High level support

Challenges
- Necessity to redirect an existing contract to support the acquisition
- Congress disapproved of
  - Reliance on sole source contracts
  - JNN's lack of status as a program of record
  - Amount of funding for a non-POR
- Perceived overlap with WIN-T
- DOT&E disapproved of the perceived lack of testing

RAND

Army Rapid Acquisition of C2   17

As evidenced in the detailed case history discussed in the companion report, JNN was a relatively complicated program. Among the cases we examined in this work, JNN encountered some unique challenges. In particular, its dollar size eventually drew the attention of both Congress and some Office of the Secretary of Defense (OSD) oversight organizations. Moreover, the rapidity of its acquisition relied on sole-source contracts, and over time Congress, the Army, and DoD staff challenged that arrangement. Additionally, DOT&E disapproved of what it perceived as a lack of testing and its avoidance of status as a program of record. By the time of these developments, of course, the program was already well into production and fielding, its value was evident to users, and JNN equipment was being authorized for all deploying forces.

JNN also benefited from a number of enabling factors. The wartime urgency to deliver beyond-line-of-sight communications to support Internet Protocol-based connectivity to the warfighter drove the institutional Army to support the acquisition. As with the other two case studies, but to an even further degree, JNN-N relied on high-level champions to overcome obstacles to securing funding and performing the acquisition, and it was possible for the system to secure this support by its demonstrated delivery to the theater of the needed capability. Supplemental funding was a cornerstone of the

31

acquisition effort. JNN also benefited from decisions by the program staff to leverage existing programs and program documentation to justify the acquisition, provide contract vehicles, and ensure sustainment in the long term.

JNN illustrates a number of lessons learned for the Army. It is clear, and also consistent with the experience of CPOF, that war drives rapid acquisition of military capabilities. In the context of war, the Army and DoD will support the flexibility necessary to perform even large-scale acquisitions rapidly, and warfighters will "take what they can get" in terms of the capability delivered, if the system is a useful improvement.

To perform rapid acquisitions, the Army may need to promote the practice of seeking related programs to leverage existing requirements documents, contracts, and sources of funding. Moreover, to enable the new capability to endure, the Army may want to consider ways to insert the new capability into an existing effort. In particular, based on JNN-N's acquisition, the Army may be able to use Engineering Change Proposals on existing contracts to initiate the work on a new related system quickly.

JNN-N, along with the other programs we discuss in the case studies, demonstrates that future rapid acquisitions should make every effort to secure high-level operational champions to protect the program in its initial stages, to provide top cover for initial development and testing, and to assist with persuading the larger Army institution of the value of the capability. The existence of high-level patrons may well be essential to make the Army bend existing processes so they deliver capabilities more quickly.

Securing funding will likely be a challenge for future proposed rapid acquisitions. JNN relied on supplemental funding, as did CPOF. When supplemental funding is no longer available, the Army may need to work as an institution to provide the necessary sources of flexible, quickly taskable funds.

Another lesson learned is that the Army can sustain rapidly acquired systems, especially during an initial period, by relying on multiyear warranties similar to those used by JNN. If the Army desires to switch to supporting the system with Army personnel, the multiyear warranties give the institutional Army time to make the decision it needs to retain the capability, and provide the appropriate personnel with the training in sustainment procedures for the system.

Figure 2.12. Some Factors Enabling Success Are Unique to Wartime

Factors Enabling Rapid Acquisition of Army C2

		FCC	CPOF	JNN
Wartime-driven attributes	Answers need the Army perceives as valid and urgent		X	X
	Able to rely on a source of immediate flexible funding		X	X
	Built on matured technology (COTS/GOTS)	X	X	X
	Users accept less than 100% desired performance	X	X	X
	Army/DoD accept higher operational risk	X	X	X
Normal but affected by wartime	Leverages existing programs and documentation		X	X
	Demonstrates useful capability quickly	X	X	X
	Endorsed and advocated by operational champion	X	X	X
Common to successful acquisition programs	Coevolves the technology and the operational concept		X	
	Plans carefully a transition to normal acquisition status to ensure long term existence and support		X	X
	Includes ongoing user feedback during development	X	X	X
	Ensures personnel on program staff consistent			X
	Ensures contractor personnel consistent	X	X	X
	Retains senior retired military consultants as advisors		X	

RAND

Army Rapid Acquisition of C2 18

The table in Figure 2.12 lists the main factors we identified that enable a rapid acquisition process to be successful. The top section of the table shows factors we believe to be particular to a wartime context. The middle section shows factors that, while not unique to wartime, were significantly affected by the wars. The bottom section shows factors we believe are common to successful acquisition programs, regardless of their rapidness or the existence of war. Not all of these factors may have equal importance in all rapid acquisitions, but future programs attempting rapid acquisition of C2 or other IT-based systems may want to consider assessing their plans against this list.

Clear in this figure is the distinction between FCC and the other two case study programs. The acquisition of FCC incorporated fewer of the factors listed here. More so that the other programs, it lacked a mission the Army perceived as urgent, a flexible immediate source of funding, and as of the writing of this report (2009–2010), we saw no evidence of a transition plan. In addition, FCC did not, as JNN-N and CPOF had done, leverage contracts, documents, or staff from any related program such as IBCS. As of the writing of this report, FCC had not successfully transitioned to an enduring capability, and it had struggled to find a mechanism to do so. This difference may show that even the omission of a subset of the factors on this list by program staff may cause difficulties for the program to persist in the long term.

If we make a judgment based on this list, it is that it may be quite hard for the Army to successfully perform rapid acquisition outside of a wartime context. The top eight factors on this list all seem critically important, and driven or affected by wartime. In particular, rapidity will require the Army to perceive the need as valid and urgent and for the Army (and firstly, Congress) to provide a immediate flexible source of funding. These two items may be particularly difficult to achieve without war motivating the necessary organizational and legislative agreements.

While the Army does not control whether war exists, if it wants to institutionalize rapid acquisition as a peacetime function,[18] replicating many of the aspects of the wartime environment will be required, at least to some extent. The Army may want to consider setting up structures to encourage and protect the factors over which it can have some control. For instance, the Army can institute policies that provide Program Managers with flexibility in prioritizing and delivering on user requirements (thus requiring requirements developers and decision makers to accept development and delivery of capabilities less than the desired 100 percent solution, and in a way not entirely prespecified by the user. The Army could ensure delivery of a useful capability by requiring the program to make those tradeoffs in continuous feedback with actual system users (i.e., not just users as represented by a skilled TRADOC Capabilities Manager or a validated requirements document). While not appropriate for all acquisitions, for rapid acquisitions the Army can insist additionally on the use of mature technology (COTS/GOTS). These changes to the traditional peacetime acquisition process would improve the speed at which new capability is developed and fielded.

---

[18] We anticipate that performing rapid acquisition will be difficult without an environment of urgency, whether from war or another known security challenge (for instance, a known cyber compromise of a major defense system). We are not aware of any formal analysis that would support institutionalizing rapid acquisition processes as a peacetime function. However, DoD has indicated it intends to do so. A June 14, 2012 DepSecDef Memo on "Rapidly Fulfilling Combatant Commander Urgent Operational Needs" directs the department to establish policy and procedures to conduct rapid acquisition. Since most of these processes were created ad hoc, DoD seems to have assumed that institutionalization (the opposite of ad hoc) is part of the solution. However, there are costs as well as benefits to institutionalizing rapid acquisition that should be more fully explored.

Figure 2.13. War Drives Many Factors Enabling Rapid Acquisition, and the Enabling Factors Influence Each Other

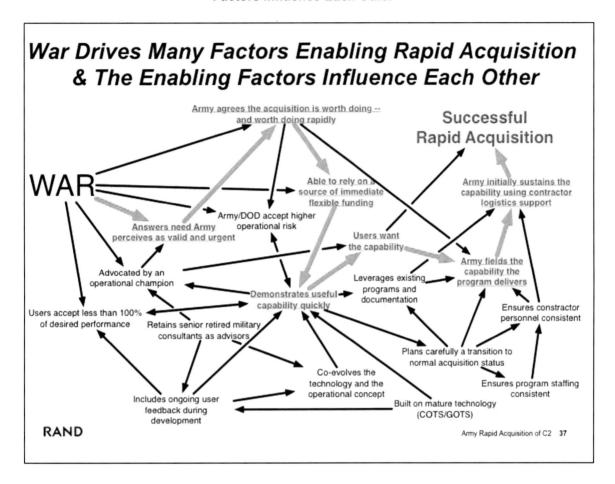

Figure 2.13 shows how the key factors identified in Figure 2.12 lead to success in rapid acquisition of C2 systems.[19] Understandably, this figure is likely to seem complex to the reader, and we ask for the reader's patience and investment in understanding it because we believe it makes two very important points, and moreover that it can serve as a guide to future program managers wanting to do rapid acquisition. First, clearly war drives much of rapid acquisition. Second, the enabling factors do not exist independently, but depend on each other and influence the likelihood of each other existing, making up a sort of "ecological community" of rapid acquisition enabling factors. We believe that the more of these factors that are present in a given rapid acquisition activity, the more likely the acquisition is to be "successful."

Notably, war drives six of these factors. In particular, war engenders needs, which personnel within the execution, oversight, and funding apparatuses of the Army will

---

[19] There are slight distinctions in the designation of "war-driven" between this Figure (2.13) and the table in Figure 2.12 because, for this figure, we insisted on more direct causality.

perceive as valid and urgent. If those proposing development of the capability convince the Army the acquisition is "worth doing" and also "worth doing rapidly," the nascent program will likely secure immediate funding from a flexible source, provided Congress has made any viable funds available. If the need is perceived as valid, in particular after the program demonstrates a useful capability, the program can and should seek a high-level operational champion within the Army to advocate for the need, existence, funding, and usefulness of the capability, and to assist it in overcoming obstacles. In addition, CPOF found it useful to retain senior retired military officers to help both with securing access to the desired system champions and with developing the operational concept and refining the technology during development.[20]

War not only convinces the institutional Army there is a need; it also persuades the Army and users to accept additional risks and performance compromises. For rapid delivery, the Army almost inevitably must endure additional reliability, maintainability, security, and interoperability risks. Also, users of the system will have to satisfy themselves with the capabilities that are deliverable in an immediate time frame. There is essentially no time available when conducting a rapid acquisition to mature technology, and there are limits as well to the achievable degree of vetting and system integration.

By relying on mature technology and including users early on and throughout development, programs can "coevolve" the technologies, system integration, and operational concept to prioritize requirements and develop any new ones that may be necessary to satisfy users. Whether the requirements have been formally validated in every detail is less informative for the design of the capability than the direct feedback of users who will use the system when it is fielded.

Having a clear need, a source of funds, and evolving the capability with user feedback will conceivably enable a program to demonstrate a useful capability quickly. When the capability can prove to potential users and high-level patrons that it will be useful, as users learn about the capability, they will start to want it. In particular, the system's operational champion can assist with connecting the system with potential users and providing support and an organizational venue for testing.

Even if users want the capability, the program must still arrange fielding and sustainment of the system. One of the major lessons learned from the case studies is that rapid acquisitions should seek to leverage existing related programs. Leveraging existing programs can help when initiating the program, by supplying to the effort existing contracts which can then be modified to deliver the new capability. A related program can also provide "a home" for a rapid acquisition to sustain itself in the long term. The

---

[20] The use of senior retired military as consultants in theater became a controversial issue. Such former military officers are useful to the contractors developing new equipment in terms of both advice and feedback from a user (warfighter) perspective, and also in terms of marketing the product. Some observers have noted a potential conflict of interest when the retired military is under contract to the firm developing or producing a particular item.

Army inserted two of the programs we examined here in other larger programs. In one case, the rapid acquisition became the initial increment of capability for the program into which it was inserted. In the other case, the rapid acquisition outperformed the initial development activity and "swallowed" the larger program. In the second case, while the outcome may be challenging for program personnel to accept, the users have the benefit of the best capability available, and the Army is well-served by the result. In both of these cases, the Army ensured competent immediate support for the new capabilities using contractor logistics support, even if the Army desired later to transition the support functions inside the Army to reduce costs. The initial contractor logistics support granted the institutional Army more time to train its own personnel to take over more of the sustainment functions over time.

In the intermediate and long term, the institutional Army will need to have more of its traditional structures in place to field and support the system. Transition planning helps rapid acquisitions persist in the force successfully. Rapid acquisitions will likely result in a number of system attributes that will have to be "backfilled," and program staff will have to staff not only the acquisition itself, but the generation of certifications, training documents, sustainment plans, further testing, and security assessments on behalf of the system. A mechanism to ensure that the structures and personnel to perform these functions is insertion into a related program. In some cases, the system may have to be redesigned or re-engineered in an improved version 2, 3, or 4, to be fielded in later stages with the necessary fixes. For efficiency in initial delivery and also consistent performance in the field, it is important to plan a transition of the program to normal status in a way that ensures, to the extent possible, stable program office staffing and, especially, contractor personnel.

This list of factors is consistent with the findings of the Defense Science Board (DSB) and Government Accountability Office (GAO) reports on urgent needs.[21] According to the DSB report:

> The most formidable barrier to rapid and effective solutions to urgent needs is available, dedicated, flexible funds. This was the primary issue raised by every witness before the task force (DSB, p. 28).

For the past decade, wartime supplemental funding has provided a source of funds for non-PORs that did not compete with established programs. The Army will soon have to face how to do rapid acquisition without it.

---

[21] Defense Science Board (DSB), *Fulfillment of Urgent Operational Needs*, July 2009. Government Accountability Office, *Warfighter Support: Improvements to DOD's Urgent Needs Processes Would Enhance Oversight and Expedite Efforts to Meet Critical Warfighter Needs*, GAO-10-460, April 2010.

(This page is intentionally blank)

3. Urgent Need, Rapid Acquisition, and Transition Processes

Figure 3.1. Introducing the Survey of Rapid Acquisition Processes

Outline

- **Introduction**

- **Case Studies**

> **• Survey of Urgent Need, Rapid Acquisition, and Transition Processes**

- **Conclusions and Recommendations**

RAND

Army Rapid Acquisition of C2 **19**

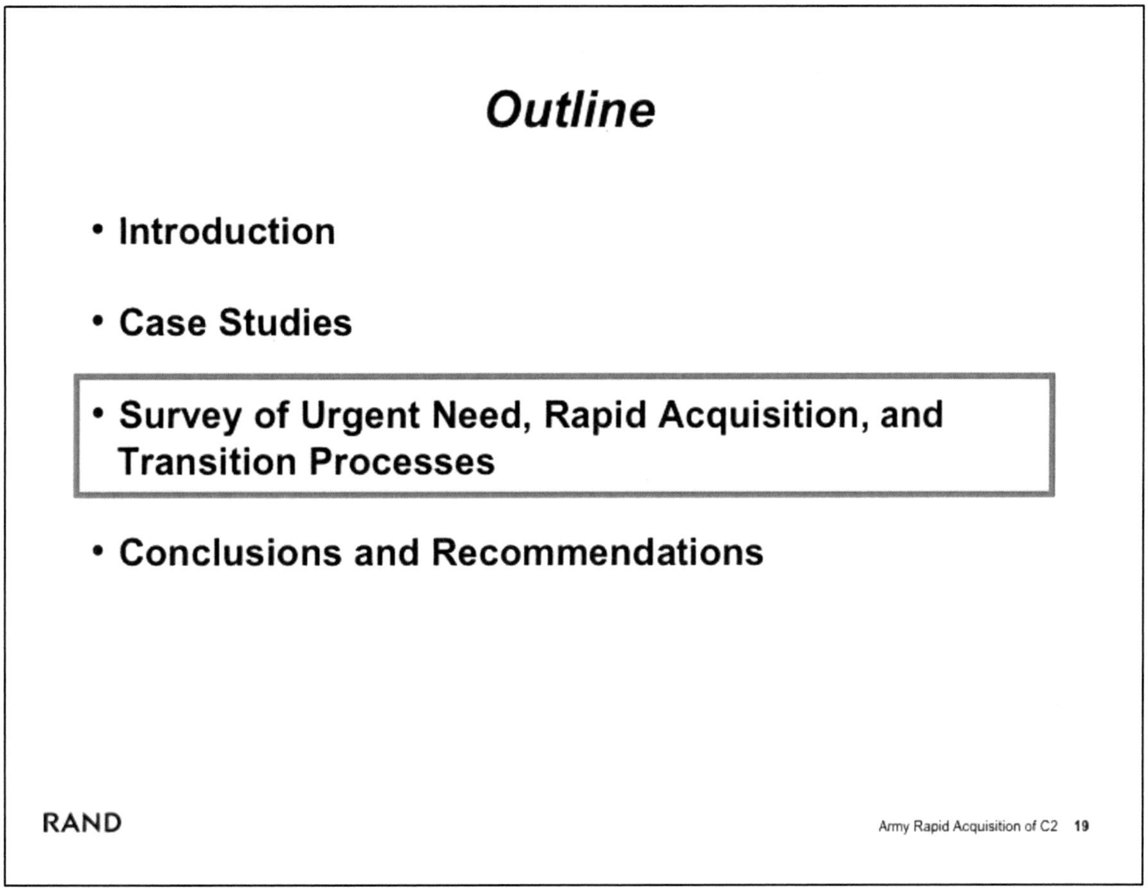

While the primary focus of our work was analysis of the three case study programs, to investigate three issues we reviewed a sizable sample of the existing rapid acquisition processes DoD and the Army have instituted:

1. What, if anything, in those processes supports or might support future rapid acquisitions of C2 systems?
2. What, if any, was the relationship of the case study acquisitions to these formally-designated rapid processes?
3. What lessons can we learn from these processes, and how might they inform recommendations regarding rapid acquisition of C2 for the Army?

This chapter broadly describes three sets of processes:

- Processes that identify and validate urgent operational needs;
- Rapid acquisition processes that acquire and deploy solutions to the capability gaps identified in the urgent needs processes; and
- Transition processes that determine whether and how the rapidly acquired systems fit into force structure over the long-term.

The three sets of processes contain elements of requirements, budgeting, and acquisition processes (where acquisition includes contracting, test, and sustainment functional areas, among others), paralleling the functions of the corresponding traditional processes. This review also helps identify factors affecting rapid acquisition programs, which provides an analytical framework we apply in the case studies. More detailed descriptions of the processes and more citations supporting this analysis reside in Appendix A of this document.

Figure 3.2. Three Sets of Processes Are of Interest[22]

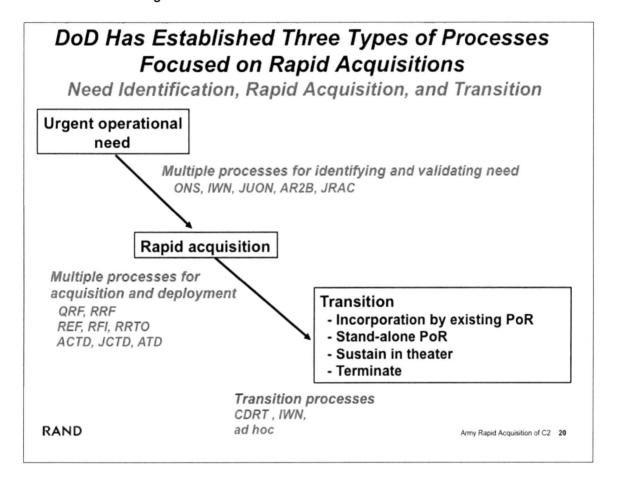

Figure 3.2 shows three broad categories of processes designed to aid DoD or the Army with rapid acquisitions, broken out by the stage at which they are designed to help.[23] In essence, for defense acquisitions to occur rapidly and persist in the force, as we discussed indirectly with the case studies in Chapter 2, all three categories of activities (requirements development, acquisition, and transition to a structure supporting an enduring capability) must occur, regardless of whether they occur through one of these named processes, via a standard process, or by an ad hoc method. The figure provides a simple illustration of the relationship among urgent need, rapid acquisition, and transition

---

[22] Acronyms shown in the figure represent the key processes of interest in this study and are defined in the Glossary and in Figure 3.3.

[23] In theory, there should exist a fourth set of processes designed to expedite testing. However, we know of no such designated testing processes designed to support rapid acquisition, and we did not have time to pursue that topic sufficiently to confirm its validity. To our knowledge the only candidate to represent this type of formalized rapid testing for C2 might be the twice-annual Army Network Integration Evaluation (NIE).

processes. Conceptually, they are sequential, reflecting different steps or phases: first identify and validate the operational need (capability gap); then develop/acquire a solution (or system) addressing that need, and deploy it; and, finally, determine the longer-term status of the newly acquired system. The first two steps are driven by the urgency of the warfighter. The third step (transition) is not necessarily urgent; rather, the decision on what to do with the new system must account for the longer-term modernization strategy. However, it is a process that uniquely needs to occur for non-PORs, which are frequently urgent acquisitions.

The urgent need processes identify and validate capability gaps, usually generated by operational units in theater or those preparing to deploy. Urgent operational need (UON) identification is equivalent to an abbreviated requirements process. These needs (or capability gaps) are generally identified by commanders of deployed or deploying forces, then validated by combatant commanders (COCOMs), and finally submitted to either Joint- or service-led rapid acquisition processes to procure a solution. An example from the case studies in Chapter 2 was the validation of an operational needs statement for the 3ID by the Battle Command General Officer Steering Committee that drove the development of JNN-N.

The urgent needs phase begins with the submission of an identified capability gap to command authorities and ends with the assignment of responsibility to a specific service or organization. The main difference between urgent needs and the traditional requirements process is that urgent needs processes are more narrowly focused on a specific operational unit, rather than a mission area or the larger force structure.

There are two kinds of urgent operational needs. The first reflects a lack of existing equipment required to carry out a unit's mission. In this case, the need for the equipment (systems) is already recognized and the solution already exists; it is just that a particular deployed or deploying unit does not have that equipment (or enough of those systems) to carry out the mission. This is a common issue for units deploying outside their normal mission area—for instance, an artillery battalion deploying as an infantry battalion. The second kind of need is the result of an emerging threat (such as IEDs) or operational experience (such as the need for mobile communications or tracking friendly forces); it is this latter kind of capability gap of interest here. Funding of urgent needs, the rough equivalent of the traditional budget process, is usually part of either the need identification phase or the rapid acquisition phase, depending on the specific processes. This includes prioritizing needs and allocating resources against those needs.

Rapid acquisition includes activities such as developing or finding a solution, contracting, program management, limited testing, and field support. This is roughly equivalent to the functions inherent in traditional acquisition processes. Rapid acquisition processes begin with the validated urgent operational need and end with the fielding and support of a system to meet that need.

The systems acquired through nontraditional means are generally not PORs. The transition decision is the point at which a determination is made about what to do with the newly acquired system now that it has been fielded in theater. As we will discuss further in the section on Capabilities Development for Rapid Transition (CDRT), there are four distinct possibilities. The first two are to make the non-POR into a POR, either as a stand-alone program or by incorporating it into an existing or future POR. (From our prior discussion, the Army thus incorporated both JNN-N and CPOF into existing programs.) Thirdly, sustainment in theater means that the Army[24] has decided to treat the non-POR system as if it were a POR with limited use in theater. This would include more centralized funding and sustainment activities, relative to the "every unit on its own" that would exist prior. Lastly, programs can be terminated, which in practice really means that the system can still be used in theater by the acquiring operational unit, but the unit will need to fund sustainment out of pocket, and the non-POR systems themselves would not become part of the standard equipment package for the unit.

It is important to note that the urgent need and rapid acquisition processes have a fairly specific and precisely defined goal: meet urgent warfighter needs as quickly as possible by equipping the requesting unit with a satisfactory solution. There is usually little or no consideration of fielding such systems more widely to the entire Army; nor are lifecycle sustainment issues usually considered. The JNN and CPOF experiences support this assessment.

---

[24] The decision is made at a senior level in the Army, for instance by the CDRT board we discuss later in this chapter.

Figure 3.3. Process Interactions and Interdependencies Can Be Relatively Complex

Process Interactions and Interdependencies Can Be Relatively Complex

Name of Process	Joint/Army	Purpose
Operational Needs Statement (ONS)	Army	Identify UONs
Immediate Warfighter Need (IWN)	Joint	Identify UONs
Joint Urgent Operational Need (JUON)	Joint	Identify UONs
Army Requirements and Resourcing Board (AR2B)	Army	Identify and Prioritize UONs
Joint Rapid Acquisition Cell (JRAC)	Joint	Identify and Prioritize UONs
Quick Reaction Fund (QRF)	Joint	Fund UONs
Rapid Reaction Fund (RRF)	Joint	Fund UONs
Rapid Equipping Force (REF)	Army	Identify and Procure UONs
Rapid Fielding Initiative (RFI)	Army	Procure UONs
Joint Improvised Explosive Devices Defeat Organization (JIEDDO)	Joint	Procure UONs
Mine Resistant Ambush Protected (vehicle) Task Force (MRAP TF)	Joint	Procure UONs
Rapid Reaction Technology Office (RRTO)	Joint	Procure UONs
Capabilities Development for Rapid Transition (CDRT)	Army	Transition UONs
Immediate Warfighter Needs (IWN) Process	Army	Transition UONs

RAND   Source: "Capabilities Development and System Acquisition Management: 2010 Executive Primer," Army Force Management School (AFMS), May 2010 and April 2009; and DSB report "Fulfillment of Urgent Oeprational Needs," July 2009   Army Rapid Acquisition of C2   **29**

The processes listed in Figure 3.3 are representative of the many Army and Joint rapid acquisition processes as of 2010 (aside from Special Operations Command's own acquisition and rapid acquisition processes supporting special operations forces).[25] The chart lists different types of efforts. Some are organizations (JRAC, REF, RRTO), some are processes (ONS, JUON, CDRT, IWN), some are programs (RFI), and some are all three (JIEDDO, MRAP). Appendix A describes these processes and organizations in more detail. There is some degree of overlap among these processes and organizations in terms of their functions and area of responsibility. No single process or organization has the roles, responsibilities, and authorities to move an urgent operational need through identification, validation, rapid acquisition, fielding, support, and transition. This means

---

[25] A GAO report identified 31 distinct processes/organizations associated with urgent needs and rapid acquisition. See Government Accountability Office, *Warfighter Support: DOD's Urgent Needs Processes Need a More Comprehensive Approach and Evaluation for Potential Consolidation*, GAO-11-273, March 2011.

that two or more organizations and associated processes are necessary to complete the lifecycle of a rapidly acquired system.

The final column in Figure 3.3 shows four sets of purposes: identifying a need, funding a need, procuring a solution, and transitioning the resulting non-POR. In practice, the distinction among these functions is not always clean. For instance, organizations that identify and fund UONs are generally associated with a processes to procure a solution. Both the AR2B and JRAC address funding issues as part of their respective prioritization functions. JIEDDO, MRAP, and RRTO have separate funding lines in either the basic DoD budget or the supplemental (contingency) fund.

There are established relationships among some of these processes and organizations. JRAC is the organization that manages the JUON process. In the Army, the ONS, AR2B, and REF are processes or organizations that identify and validate needs, prioritize and fund those needs, and develop or procure a solution, respectively. Of the items listed, JNN-N used one, the ONS, and CPOF used one, CDRT. (We will discuss CDRT shortly.)

JIEDDO and MRAP are the examples most studied, although they were not focused on developing C2 systems. But because these two are well studied, they may offer lessons for other nontraditional processes. Both make extensive use of existing mature technology and are funded with the wartime supplemental budget. Both also have high visibility and consistent senior-level support within OSD and the services. These factors turn out to be critical to the C2 programs we examined, as well as to the successful functioning of rapid acquisition processes in general. One difference between these two organizations and their associated processes and the other organizations or processes in Figure 3.2 is that they are relatively narrowly focused. For instance, JIEDDO is focused on a particular threat (improvised explosive devices) that has far-reaching implications and involves a broad range of potential mitigating actions or systems, while the MRAP TF is focused on procurement of a specific class of vehicles with specific attributes (resistance to and dissipation of damage inflicted on military vehicles from roadside IEDs).

In the middle of the table, REF and RFI illustrate the difference between "equipping" and "fielding." The REF finds specific solutions to specific needs for a specific operational unit, with a goal of fielding the capability within 90 days of urgent need approval.[26] The RFI organization is responsible for ensuring that every soldier deploys with an up-to-date, basic set of equipment and operates within the expected deployment timeframe for the unit.[27] That responsibility and associated products have recently been distributed throughout PEO Soldier. The planning responsibilities of RFI have been

---

[26] Scott Stearns, "The Rapid Equipping Force: Supporting the American Warfighter," *Infantry Bugler*, National Infantry Association, Fall 2008.

[27] Robert A. Rasch, "Lessons Learned from Rapid Acquisition: Better, Faster, Cheaper?" Strategy Research Project, U.S. Army War College, Carlisle Barracks, PA, December 1, 2011.

moved to a new element on the PEO staff. RFI thus illustrates that traditional Army organizations play important roles in rapid acquisition processes.[28] To our knowledge, none of the programs we examined in the case studies received support from either of these organizations.

---

[28] See Jason Sherman, "Reorganized PEO Soldier to Improve Focus on Body Armor, equipment," *Inside the Army*, November 30, 2009.

**Figure 3.4. Recent DSB and GAO Reports on Urgent Needs
Processes Identify Similar Sets of Challenges**

Recent DSB and GAO Reports on Urgent Needs Processes Identify Similar Sets of Challenges

DSB (July 2009)

- Lack of management data

- Limited feedback from users

- Risk adverse acquisition culture

- Ad hoc nature of the processes
  - Coordination issues
  - Little institutionalization

- Differentiating responses for different classes of needs

- Dedicated, available, flexible funding

GAO (April 2010)

- Lacks management framework
  - Data collection
  - Feedback from users

- Can't assess process effectiveness or efficiency

- Guidance is fragmented and outdated
  - Roles and responsibilities
  - No standardized responses

- Limited training

- Funding availability

RAND

Army C2 final brief 7 11 August 2010

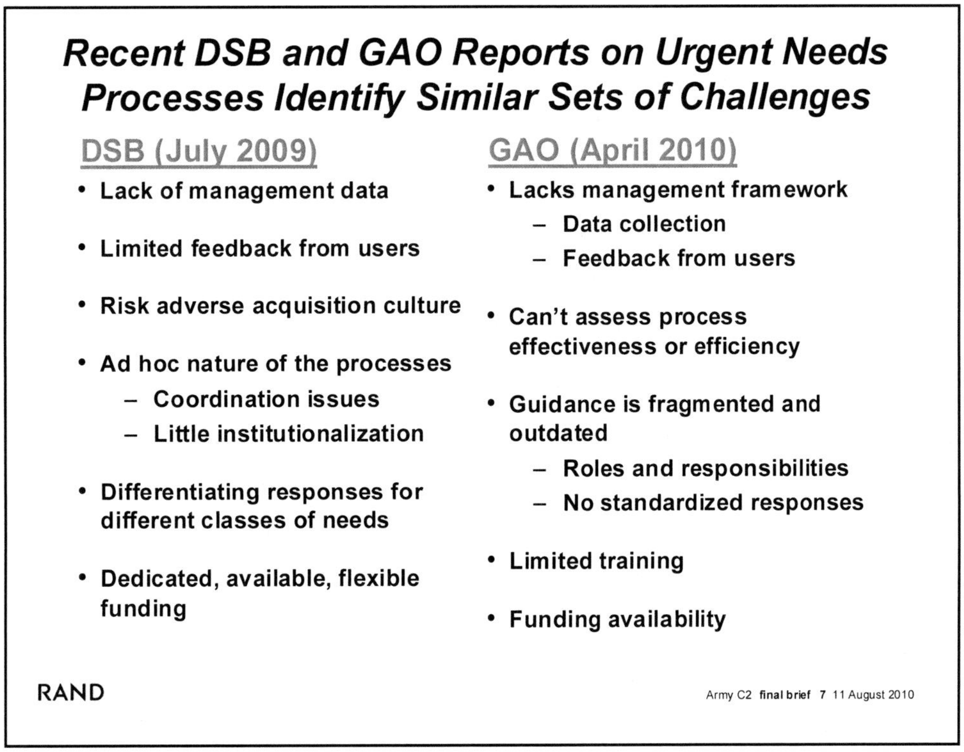

In studies conducted roughly concurrent to this one, both the GAO and DSB examined a wide range of rapid acquisition processes and organizations.[29] Their basic findings are highlighted in Figure 3.4. Interestingly, the two studies reached very similar conclusions: rapid acquisition processes pose significant issues and challenges to DoD, both individually and collectively. These issues include:

- Policies, processes, and organizations are fragmented, with little or no central coordination.

- There is potential for conflict among these disparate processes and organizations due to differences in areas of responsibilities and timing.

- The roles, responsibilities, and authorities are not fully specified in official policy documents.

- There is no single senior official, below the Secretary of Defense, who has the responsibility and the authority to go from a validated urgent need, through fielding a solution, to deciding what should be done with that system in the long run.

---

[29] DSB, July 2009; GAO, April 2010.

- There are no metrics or formal feedback mechanisms allowing DoD to assess both the effectiveness of rapid acquisition processes and how well rapidly acquired systems satisfy user needs.
- There are few formal transition processes to determine the long-term status of the fielded system.

The DSB report highlights a lack of data collection that would enable process managers to track an urgent need and its solution through the process, from origination to fielding. Officials are therefore unable to determine how long it takes to respond to an urgent need and whether that need was satisfied. The DSB noted that the traditional "deliberate" acquisition process is generally risk averse, whereas a rapid response may entail accepting certain risks. The ad hoc nature of the many processes, some established for a narrowly defined need and others established with a broader charter, leads directly to coordination problems across agencies, with little or no attention given to institutionalizing these processes by capturing lessons from recent experience. The DSB also noted that there was no up-front process to direct different classes of needs and solutions down the appropriate path. For instance, the response to a validated need for more of an existing POR should be treated differently from the response to a need for an entirely new capability. Lastly, DSB recognized the need for flexible funding to satisfy needs, something that is likely to be an increasing problem as budgets decline and the military transitions back to a more peacetime posture.

The GAO report (GAO-10-460) documents similar challenges and problems with existing urgent needs processes in both DoD and the Army. GAO notes the absence of a sufficient management framework for urgent needs processes, including operating procedures for agencies, collection of information that enables program management and oversight, and a mechanism for obtaining feedback from users. As a result, the Army and DoD more generally can assess neither how well the processes are performing nor how well the fielded systems address user needs. While the memos establishing the various processes contained some guidance on goals for the process, how the process should work, and the roles of stakeholders, many processes have changed over time as a result of experience gained and changing circumstances; the official guidance has not been updated. In addition, standard responses have not been established for different sets (categories) of solutions, resulting in inconsistency in how similar needs are addressed. GAO also notes that there is limited training for officials at all levels as to how the process should work and what their specific roles should be. Funding availability has also been an issue in some cases, in part because OSD has not identified an official with primary responsibility for making funding decisions.

Both reports mention that the rapid acquisition authority Congress explicitly granted DoD has been used infrequently.[30] This authority, established in the FY2005 National Defense Authorization Act, provides the Secretary of Defense the authority to reprogram funds (up to $100 million annually) to meet an urgent need associated with fatalities (several years later Congress changed this last criterion from fatalities to casualties). GAO notes that the authority has been used only four times (as of 2010) and attributes this, in part, to the lack of a single senior official with primary responsibility for engaging and operating DoD's urgent need and rapid acquisition authorities and processes.

In addition to the GAO and DSB reports, a recent Army Science Board (ASB) study assesses the urgent needs and rapid acquisition organizations they examined as successful.[31] The success was enabled in part by senior leader support and the ability to bypass normal processes. However, the study notes that these ad hoc organizations (or processes) also face challenges, including ensuring a stable funding stream when the supplemental budget disappears, limited lifecycle planning and transition planning, and some duplication and redundancy. These challenges are the same that we found in our three case studies and our review of rapid acquisition and associated processes more generally.

Our review of DoD urgent operational needs and rapid acquisition policies and processes generally agrees with the conclusions from these DSB, GAO, and ASB reports.

---

[30] Although we ourselves have not directly studied why, it might be reasonable to hypothesize that the reason this rapid acquisition authority has been used infrequently is that the criteria to use it required a determination that lack of the capability would lead to loss of life, the dollar cutoff ($100 million) for an eligible program was too low to fund any broadly needed capability, and the decision-making authority was reserved at too high of a level (Secretary of Defense, undelegatable) to be a useful mechanism for program managers, especially for such a low-dollar-amount decision. In other words, anything DoD needed badly enough to make it to the desk of the Secretary of Defense would likely have cost more than $100 million in a given year and, furthermore, may not have been tied directly to loss of life—perhaps just loss of combat effectiveness. For instance, JNN-N was criticized for not using this authority, yet its high cost per year would have made it ineligible, even if the lack of it could have been tied to loss of life or casualties. See Defense Acquisition University, Acquisition Community Connection website, "Rapid Acquisition Authority to Respond to Combat Emergencies," January 14, 2005.

[31] See Majorie Censer, "Army Science Board Recommends DoD Establish Innovation Cell," *Inside the Army*, February 1, 2010.

Figure 3.5. Urgent Need and Rapid Acquisition of C2 Systems Face Many Challenges

---

Urgent Need and Rapid Acquisition of C2 Systems Face Many Challenges

- **Fulfilling immediate needs vs. acquiring capability for longer-term**
  - **"equipping" vs "fielding" may be an important distinction**

- **Identifying, tracking, coordination, duplication**

- **Integration and interoperability**

- **Test and certification requirements**

- **Short vs. long term funding**

- **Training and sustainment**

RAND

Army C2 **final brief** 8 11 August 2010

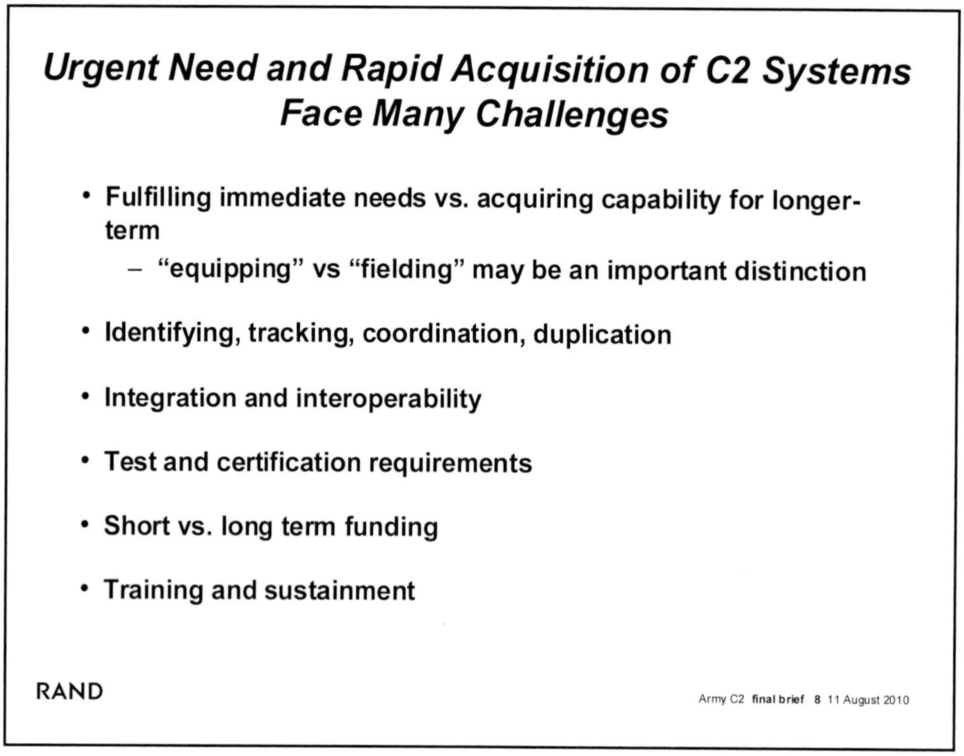

In addition to the issues identified in the DSB and GAO reports, there are other challenges associated with urgent need and rapid acquisition processes. Figure 3.5 lists some of these issues, which we have identified in both the review of the processes and in the case studies. At their core, these processes introduce a tension between the need to respond to immediate operational needs of deployed (or deploying) forces and the longer-term need to acquire capability that allows DoD and the Army to maintain or increase its advantages over a wide range of potential adversaries. The other challenges come after a decision has been made to acquire a specific capability rapidly.

Once fielded, there is no formal system to identify and track which operational units have which non-POR systems. As a result, commanders in theater, as well as the parent military service, may not know how many non-POR systems are fielded and which units have them, which can complicate operations and sustainment in theater. Given the lack of coordination at the beginning of the rapid acquisition process and the fact that each operational unit may be treated independently, there is the potential for duplication of capabilities, i.e., two different units may acquire different tactical radios or nonlethal weapons to address the same capability gap.

Integration and interoperability are potential challenges, especially for C2 systems. Limited testing prior to deployment of non-POR systems introduces operational risks in terms of both effectiveness (will the system work as expected) and information assurance and related IT certification issues. This is an important inherent tradeoff in rapid acquisition, particularly for IT-based systems. The traditional acquisition process includes substantial testing in various forums to ensure interoperability and information assurance of IT-based systems; there is an inherent bias toward taking the time to ensure that everything works as specified. Rapid acquisition processes have been biased toward meeting an operational need as fast as possible and have not included the same extensive testing regimes.

Since most non-POR systems are acquired to meet a UON in theater, funding has come from the supplemental budget. This may work in the short term, but longer-term funding then remains an issue that would need to be dealt with as part of a transition decision. Similarly, training and sustainment may be cobbled together to support the warfighter, but how these important functions are handled in the longer term is not usually considered as part of a rapid acquisition process.

Figure 3.6. IT System Acquisition Presents Its Own Challenges

IT System Acquisition Presents Its Own Challenges

- Inability to fully specify requirements up front

- Critical importance of user feedback in all phases of development and operational use

- Fast pace of change in hardware and software
  - Introduction of improved capability or new functionality
  - Shorter development timelines

- Predominance of commercial systems

Characteristics of successful C2 rapid acquisition are similar to suggested elements of a peacetime IT system acquisition process (see NRC study)

RAND

Army C2 **final brief 9** 11 August 2010

C2 systems are inherently IT systems, and the acquisition of IT systems presents unique acquisition-related challenges due to the characteristics of IT systems more broadly (see Figure 3.6). For instance, in many IT systems, especially those that are software heavy and require user feedback to develop, the full set of requirements cannot be specified up front. Requirements, which reflect desired capabilities, tend to evolve as the user gains experience with the system. This means that some form of incremental (or spiral) development and fielding will be necessary, and those increments will most likely be relatively short given the fast turnover of both hardware and software. Shorter development timelines require the use of more mature technology and, together with the fact that the commercial sector leads the military sector in IT-based technological advances for most systems, the use of COTS systems may dominate.

Interestingly, the nature of rapid acquisition mitigates these IT-unique challenges to some extent. Acquiring systems rapidly alleviates the effects of the fast pace of change in commercial-sector hardware and software. Changes in hardware and software can incorporate new capabilities identified by both the system developer and users, allowing for a natural evolution in requirements.

Smaller focused C2 or IT system acquisition programs can be more successful; they tend not to get bogged down like many large IT programs.[32] This is due in part to relatively more incremental and defined sets of capabilities.

C2 systems are a subset of IT-based systems. Many of the characteristics of successful rapid acquisition of C2 systems are similar to the suggested elements of a peacetime IT system acquisition process.[33] However, as noted in Chapter 2, war has been a significant motivator for rapid acquisition of C2. It has also been the prime motivation for setting up most if not all of the processes and organizations discussed here. This suggests that there may be challenges to acquisitions, whether C2 or IT more broadly, that try in peacetime to replicate the success of war-driven rapid acquisitions.

---

[32] Perhaps this is true because less systems integration is required. See Isaac R. Porche III, Shawn McKay, Megan McKernan, Robert W. Button, Bob Murphy, Kate Giglio, and Elliot Axelband, *Rapid Acquisition and Fielding for Information Assurance and Cyber Security in the Navy*, Santa Monica, Calif.: RAND Corporation, TR-1294-NAVY, 2012.

[33] National Research Council, *Achieving Effective Acquisition of Information Technology in the Department of Defense*, Washington, D.C.: The National Academies Press, 2010.

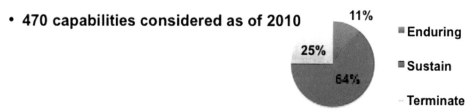

Capabilities Development for Rapid Transition (CDRT)
Transitions Some Rapid Acquisitions to Official Status

- VCSA directed TRADOC in 2004 to identify and decide what to do with non-standard equipment acquired in support of GWOT

- Both "push" and "pull" mechanisms identify candidate systems

- There are three possible outcomes for each system:
  - Enduring -- Army wide application (PoR)
  - Sustain (niche) -- supported in theater
  - Terminate -- not supported with continued funding

- 470 capabilities considered as of 2010

11%
25%
64%

■ Enduring
■ Sustain
- Terminate

Source: CDRT Information Briefing, 02 Oct 09; CDRT Info Paper, 23 Sept 2009; CDRT Update briefing Jan 2010

RAND

Army Rapid Acquisition of C2   21

As shown in Figure 3.7, the Army established a formal process in 2004 to identify non-POR (or nonstandard equipment) systems that have been fielded and to determine whether they should transition to the traditional acquisition process and, by definition, become a POR: the Capability Development for Rapid Transition (CDRT) process.[34] Interestingly, we did not know about CDRT when we began this research; nor did one of our sponsors. Another of our sponsors had only recently become aware of it.[35] This relative lack of visibility among stakeholders is an important observation that applies to

---

[34] Department of the Army, Memorandum, "Procedures for Transfer of Rapidly Equipped Initiatives/Products/Systems to Program Executive Offices (PEOs) for Life-Cycle Management," October 24, 2008 (signed by Dean G. Popps, Principal Deputy Assistant Secretary of the Army (Acquisition, Logistics and Technology).

[35] We would assume this lack of visibility into the processes is a result of insufficient publicity or immaturity of the process.

many of the rapid acquisition processes we examined: even if the processes are known in general, the detailed understanding required to use them is often lacking.

The CDRT process was established to determine whether the many promising technologies and systems procured and fielded quickly to Iraq and Afghanistan have wider applicability and deserve more permanent funding. CDRT (originally "Spiral to the Army at Large") started as an annual process. In 2008 (iteration 5), it became semiannual. In May 2009 (iteration 6), General Chiarelli, Vice Chief of Staff of the Army, made the process quarterly to accelerate the evaluation of the various capabilities. CDRT has both "push" (commander in the field submits a request to have a system reviewed) and "pull" (active search for non-POR currently in use) mechanisms.

The CDRT review process has three possible outcomes:[36]

- **Enduring**, meaning that the system or capability has wider applicability within the Army and so should become a POR. Within this category there are two types of outcomes: the system could be incorporated into an existing program of record, or it could become a stand-alone POR. In either case, it would now be subject to the traditional acquisition process and would compete for funds within the PPBES process.
- **Sustain or niche,** meaning that the system or capability has enough utility to be formally supported in theater. The non-POR would then become part of the logistics processes supporting theater requirements.
- **Terminate**, meaning that the operational unit may continue to use the system in theater, but must fund that use out of unit O&M funds. In this case, the system may not be brought back to the unit's home base in the United States.

As of January 2010, 470 capabilities had been reviewed. Of these, 51 had received the "enduring" designation, 117 had been terminated, and the remaining 302 were given the "sustain" designation. The CDRT process appears to meet the need for deciding formally what to do with the larger numbers of non-POR systems in use in theater.

Of the systems we considered in the case studies, as of the writing of this report, only CPOF had gone through CDRT. However, CDRT did not seem to be a significant contributor to the transition process in the opinion of the CPOF program staff, as they did not mention it during the interviews, while a great a deal of the discussion focused on the manner of choreographing the transition to the Army from DARPA.

The establishment of CDRT in 2004 indicated that senior leaders in the Army were aware of the potential opportunities and challenges presented by the rapid acquisition of systems to support the warfighter using nontraditional means. The evolution of the

---

[36] The CDRT decision is not necessarily final, and we believe that is a good thing, as a capability judged initially to not be worth keeping may later prove its worth.

process (from annual to semiannual to quarterly) reflects an awareness of the growing number of such non-PORs in use. However, the entire non-POR-to-POR lifecycle is not governed by any organization below the Secretary of the Army and the Chief of Staff of the Army. This lack of ownership of the complete lifecycle contributes to difficulties coordinating the many ad hoc rapid acquisition processes within the Army.

Figure 3.8. Example Programs (1)

Example Programs (1)
Used Both REF and CDRT Processes

Sample Set of Systems in both CDRT and REF		Iteration	CDRT Result
Escalation of Force (EOF) Kit		3	Niche
Escalation of Force (EOF) Kit	*Some systems appear in back-to-back iterations*	4	Niche
Green Laser - GBD-III Laser		3	Niche
Green Laser - Mini-Green Laser (B.E. Meyers) (Green Lasers: Minigreen 532M)		3	Niche
Green Laser - Mini-Green Laser (B.E. Meyers) (Green Lasers: Minigreen 532M)		4	Niche
Green Laser - XADS PD/G-105 Laser Dazzler		3	Niche
Green Laser - ZBOLT		3	Niche
Green Laser: ZBOLT	*Moved from niche to PoR in next iteration*	4	POR
MARCbot		2	Niche
MARCbot		3	Niche
PACKBOT		2	Niche
Rapid Deployment Integrated Surveillance System (RDISS)		5	Sustain
Tactical Mini UAV (TACMAV)		3	Term
Translators: Phraselator		3	Niche
Wellcam	*Modified system in subsequent iteration*	3	Niche
Wellcam - Confined Area Enhanced ISR		6	Sustain

RAND
16 programs

Army C2 **final brief 11** 11 August 2010

The rapid acquisition organizations and processes operate largely independently to satisfy UONs, though there are some established relationships, as mentioned previously (and described more fully in Appendix A). However, the transition from a non-POR to POR requires a decision that, formally, only CDRT can make within the Army.[37] As an example of the actions taken by CDRT, Figure 3.8 is a sample list of programs acquired through REF that have gone through the CDRT process.[38,39]

A few observations on this list generalize to the broader set of rapidly acquired programs and systems:

- There is a wide range of mission-specific systems apparent even on this short list, including lasers (nonlethal weapons), robots and a UAV, and several kinds of IT-based systems.

---

[37] It is true, however, that JNN-N made that transition without participating in CDRT. The decision received extensive oversight through other mechanisms.

[38] The Wellcam appears twice on the list because of a system modification.

[39] The full lists of programs that have gone through both REF and CDRT are not publically available.

- Some systems appear in back-to-back iterations of CDRT possibly because the system was modified the second time through or was subsequently placed in a different unit that submitted it back into the CDRT process. In addition, more permanent decisions on some systems are put off until the CDRT process has a better understanding of whether or not a capability will be enduring in the future.
- One system, the Green Laser ZBOLT, was reconsidered in the next iteration, resulting in a change in the transition decision from niche to POR. This suggests that either the threat changed or the system demonstrated a capability deemed enduring. This also demonstrates a CDRT decision change, not necessarily a bad thing, as decisionmakers may gain more information about a system over time.

Figure 3.9. Army's Immediate Warfighter Needs (IWN) Process

Army's Immediate Warfighter Needs (IWN) Process

- Army Asymmetric Warfare Office (AAWO) initiated process in 2007 with focus on JIEDDO initiatives

- Broadened scope to take initiatives from CDRT, REF, TRADOC, RDECOM, and other organizations

- Ensures rapidly fielded systems/technologies, not already transitioned by CDRT, are resourced for initial program management or life-cycle support to AAE or AMC
    - Vets initiatives for possible transition

- AAWO has strong role, along with ASA(ALT), AMC, RDECOM, TRADOC, and Hq G-2, G-3/5/7, G-4, G-8

- Process has been informal and has only transitioned an estimated 40 to 50 capabilities

Source: "Capabilities Development and System Acquisition Management: 2010 Executive Primer," Army Force Management School (AFMS), May 2010

RAND   Army C2 **final brief** 13 11 August 2010

Immediate Warfighter Need (IWN)[40] is a second Army transition process (Figure 3.9). The Army Asymmetric Warfare Office (AAWO) initiated the process with a focus on systems coming out of JIEDDO, but has since broadened its scope to include systems coming out of multiple rapid acquisition processes. It also appears to have picked up some systems that went through CDRT and revisited the transition and funding decisions. IWN seems a less formal process than CDRT, but it does bring together representatives of the acquisition, requirements, and user communities to make decisions. IWN has transitioned less than 50 small-sized (on a unit cost basis) systems as of May 2010.

---

[40] Some terminology confusion should be noted. The Army's IWN is a transition process. Within JRAC or the rapid acquisition law, an IWN is a specific urgent need.

Figure 3.10. Lessons Learned from the Survey of Established Rapid Acquisition Processes

Lessons Learned from the Survey of Established Rapid Acquisition Processes

- **A plethora of different rapid acquisition structures within DoD successfully deliver capabilities to address urgent needs**

- **The processes dedicated to support rapid acquisition do not meet all Army or DoD institutional needs in performing rapid acquisition**

- **While two processes handled the case study programs, the programs did not seem to rely significantly on their existence.**
  - **JNN-N used an Operational Needs Statement (ONS)**
  - **CPOF went through the Capabilities Development for Rapid Transition (CDRT) process**

- **Initial acquisitions usually did not consider manner of transition and life-cycle issues; case study or not -- often a later process arranges for transition**

- **Non-wartime interests govern the transition from non-POR to POR**

RAND Army Rapid Acquisition of C2 22

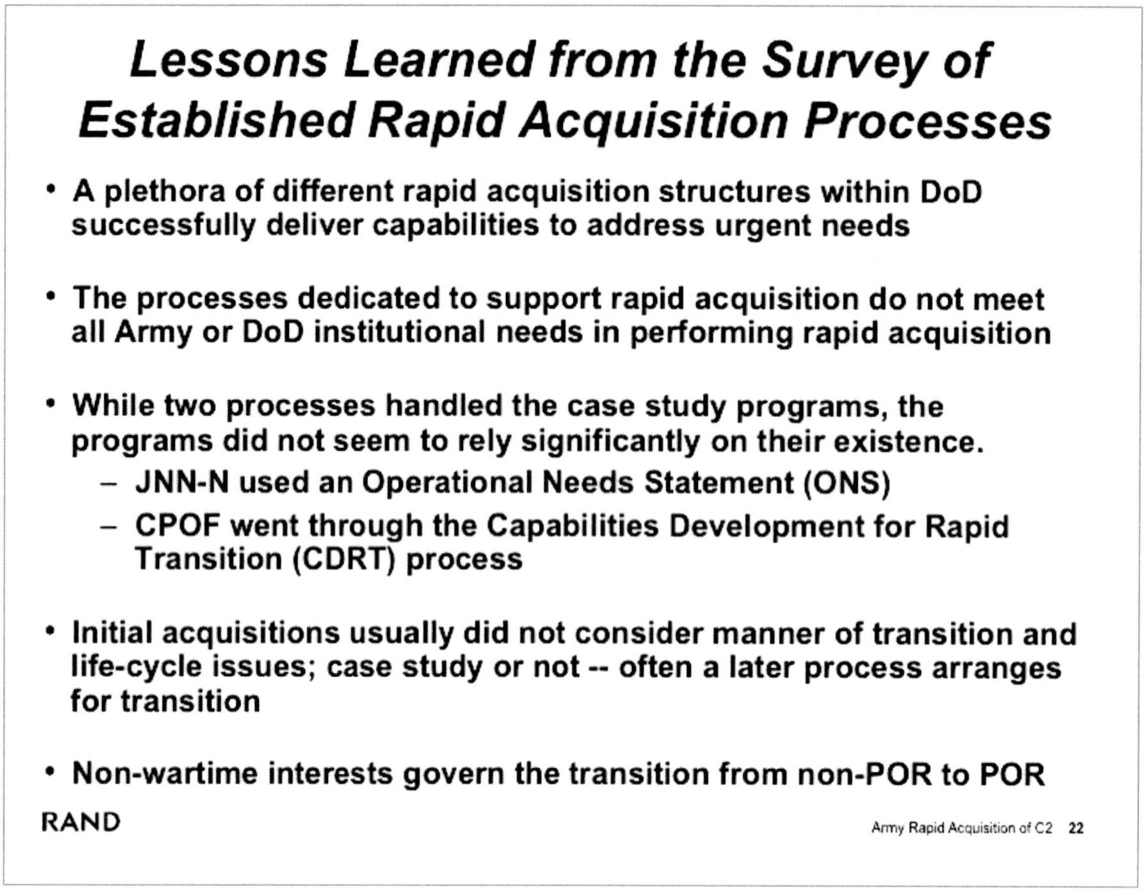

As shown in Figure 3.10, we learned a number of lessons from our review of established rapid acquisition structures in DoD. Based on the DSB and GAO reports discussed earlier, congressional language in several National Defense Authorization Acts (NDAAs), as well as discussions with select Army and OSD officials, we find general agreement that the traditional requirements, acquisition, and budgeting processes are too cumbersome to address urgent warfighter needs, and that new structures need to exist to support rapid acquisition.

However, these officially designated rapid acquisition structures have not met all of the Army's needs during the recent conflicts—our case study programs did not rely on these processes significantly. Of the three cases we examined, the interaction of these processes with the case studies was minimal. FCC, to our knowledge, did not interact with any of these structures. JNN-N had an operational needs statement validated during the early stages of the program, but also had its requirements validated in a number of other forums. CPOF went through CDRT during its transition into the Army's inventory, but the program staff did not mention CDRT as significant to CPOF's transition. It is possible that both underwent their transitions before CDRT had matured as an institutional expectation, and both programs were sufficiently high profile so as to merit the creation of special structures tailored to each.

The various processes, including CDRT, have evolved in several ways. Some processes have increased the scope of responsibility or authority, meaning they cover more areas or topics than they did initially. Some urgent needs processes have become strongly related to the rapid acquisition process that would execute the resulting program, like the Army's ONS and REF processes. As mentioned, CDRT has evolved from an annual process to a quarterly exercise.

While urgent need validation processes have been helpful in performing rapid acquisition, one of the most important lessons from this chapter is that the formal structures such as CDRT designed to transition non-POR urgent acquisitions into the standard acquisition system help the long-term sustainability of the Army's inventory. Transitioning the rapidly acquired systems is rarely considered until after a rapidly acquired solution is fielded. Transitioning enables the programs to both proactively and retroactively arrange to satisfy nonwartime interests.

Rapid acquisition processes are largely designed to address wartime needs, and have relied largely on contingency budgeting for support. The sheer number of a type of rapidly acquired systems may present a challenge as the Army winds down from a wartime footing. As such, the Army can benefit from publicizing and strengthening mechanisms like CDRT to bring the rapidly acquired capabilities that are valued by users into the standard planning process.

Figure 3.11. Institutionalizing Rapid Acquisition Poses Both Challenges and Opportunities

Institutionalizing Rapid Acquisition Poses Both Challenges and Opportunities

Challenges
- Motivating rapid acquisition in the absence of war
- Establishing funding stream
- Overcoming pushback from traditional stakeholders
  - IA
  - Testing
  - Sustainment

Opportunities
- Establishing mechanisms protecting PMs to
  - Embrace increased risk
  - Tailor acquisition strategies
- Limiting the staff learning curve resulting from ad hoc tailoring
- Strengthening useful transition processes like CDRT

* A June 14, 2012 DepSecDef Memo on "Rapidly Fulfilling Combatant Commander Urgent Operational Needs," directs the department to establish policy and procedures to conduct rapid acquisition

RAND

Army Rapid Acquisition of C2 24

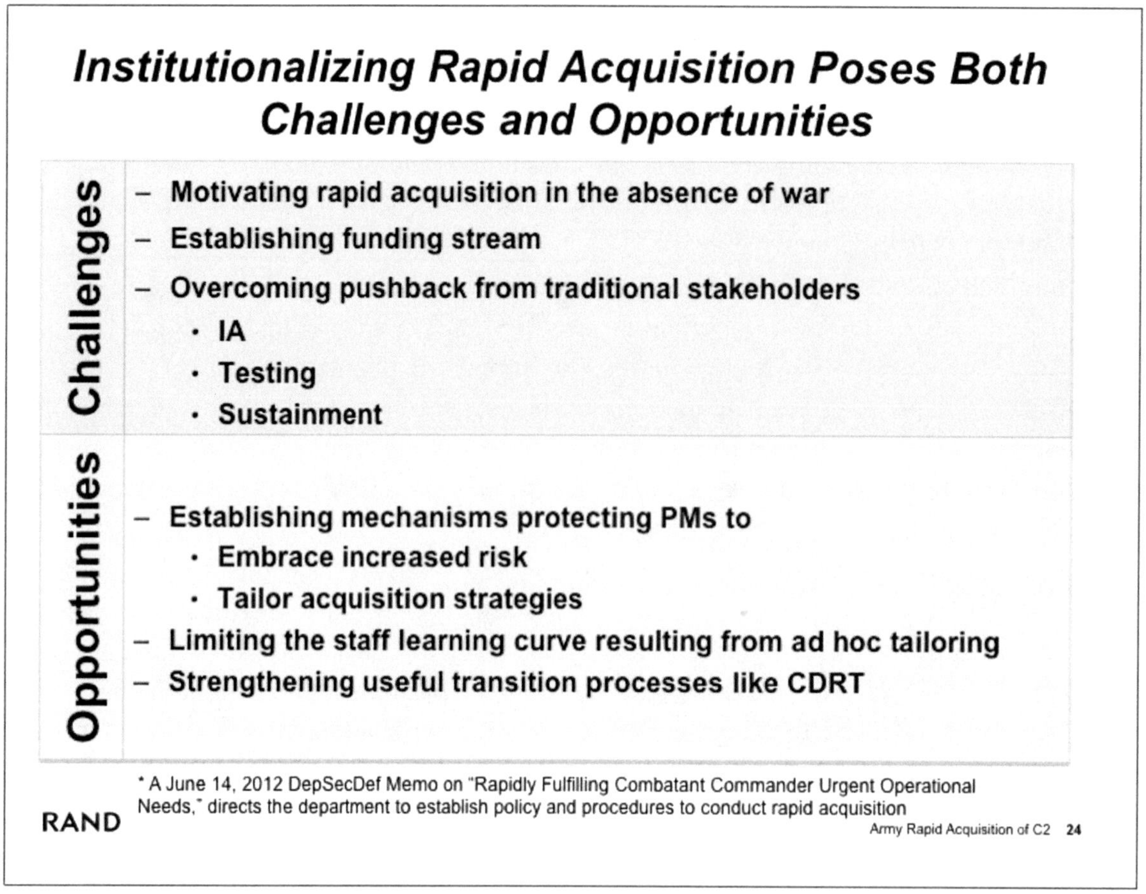

DoD has been indicating that it plans to institutionalize various aspects of the wartime rapid acquisition processes it has stood up in the past decade.[41] Figure 3.11 shows challenges and opportunities the Army may have as it goes forward with institutionalizing rapid acquisition. As indicated earlier in Figure 2.13, war itself drives many of the enabling factors for rapid acquisition. The Army may face difficulties in enabling rapid acquisition outside of the wartime environment.

The challenges the Army and DoD will likely have to overcome are how to institutionalize rapid acquisition without having it turn into standard acquisition. Stakeholders representing the concerns about information assurance (IA), the testing community, and those planning for sustainment will put significant pressure on institutional planning to increase the coverage of their topics beyond what has occurred in previous rapid acquisitions, and compared to the programs and processes we have

---

[41] As also referenced in Chapter 2, a June 14, 2012 DepSecDef Memo on "Rapidly Fulfilling Combatant Commander Urgent Operational Needs," directs the department to establish policy and procedures to conduct rapid acquisition.

discussed in this report, acquisitions acquiescing to those concerns, while enjoying lower operational risks and better reliability and maintainability, will likely deliver capabilities on a significantly longer timeline. In fact, over time, as the newly institutionalized rapid acquisition process evolves, satisfying all of these areas of interest may, in essence, bring the Army back to its normal process, at least in nonwartime.

However, the Army also should see opportunities in the push to institutionalize rapid acquisition. By establishing dedicated rapid acquisition structures and policies, it can establish mechanisms that protect program managers when they embrace increased risk or tailor their acquisition strategies in order to shorten the acquisition timeline. Also, institutionalization can preserve "lessons learned" through case studies and other analyses, limiting future staff learning curves resulting from the lack of guidance and continual reliance on ad hoc tailoring of methods. Moreover, the Army, through institutionalization of rapid acquisition, can put in place measures to strengthen useful transition processes like CDRT.

Our broad review of urgent need, rapid acquisition, and transition processes identified a set of factors that affect the operation and relative success of the processes that has informed our overall analysis and the recommendations made in this report. These factors include:

- Real urgency in the warfighter need;
- Stakeholder familiarity with the urgent need and rapid acquisition processes;
- Reliance on mature technology;
- Willingness of users to assume relatively more operational risk due to the brevity of testing and certification activities;
- Close working relationship between the organization executing the rapid acquisition and the operational user, including a feedback mechanism by which the user communicates issues to the acquirer;
- The need for rapid acquisition programs to transition to normal acquisition status as they execute.

(This page is intentionally blank)

4. Conclusions and Recommendations

Figure 4.1. Summary of Key Findings

Summary of Key Findings

High-level champions are critical to the success of rapid acquisitions

Organizational and user flexibility enables rapid acquisition of new capabilities

- Wartime environments motivate bureaucratic flexibility
- For needed equipment, with good program office communication, users will tolerate operational risk and less than 100% of desired performance
- Ongoing user feedback enables iterative capability enhancement
- Rapid acquisitions require an immediate flexibly-tasked source of funding

A large variety of named rapid acquisition processes have supplied a number of wartime capabilities for DoD and the Army

- These designated processes have not met every rapid acquisition need of the Army
- The case study programs did not rely heavily on these named processes.

Relying on existing technology and documentation speeds acquisitions

- "Faster" can mean leveraging existing requirements, contracts, and documentation from other programs
- Rapid acquisitions require mature technology

C2 rapid acquisitions have ensured field support and sustainment via contractor warranties and eventual transition to an Army POR

- C2 rapid acquisitions have relied on multi-year warranties
- Transition planning and staffing consistency during the transition is essential

RAND   Army Rapid Acquisition of C2 **25**

Here is a summary of key findings from our work. These findings came from our analysis of both the case studies and the established rapid acquisition processes. Our findings are supported both directly by this report and also by the detailed case study work reported in the companion document. Many of our findings were highlighted by the case study programs.

One of the consistent findings from the case studies was that securing a high-level operational champion for the effort was critical to the success of rapid acquisitions of C2. C2 systems, unlike individual equipment, are frequently complex and interdependent, and require broad institutional support to supply sufficient funding and overcome development, integration, and testing barriers. An operational champion helps the program in its vulnerable early stages overcome resistance to its existence (for instance, from perceived redundancy with a more-slowly-delivering program of record). He or she also connects the program with its users, by promoting its usefulness and by providing institutional support for initial testing and deployment.

Organizational and user flexibility are required for rapid acquisition, especially rapid acquisition of IT-based systems. In an environment posing an institutionally recognized urgent need, the Army has been willing to tolerate some increased operational and security risks from relying on rapidly deployed yet not fully tested systems. Even more so, in return for a needed capability, with proper communication from the program office, users will accept new systems delivering less than the desired level of performance.

Successful rapid acquisitions of C2 systems have exclusively relied on mature technology, and have frequently evolved the capability by relying on users to vote "early and often," essentially requesting feedback in the beginning and throughout development. The ongoing user feedback enables programs to iterate their capability enhancement, and it improves both the relevance and usefulness of the capability. It also provides the program office with the opportunity to socialize any performance compromises they propose to make in the rapidly delivered system.

War has been central to the performance of rapid acquisition over the past decade, and it is not apparent that rapid acquisitions can succeed without it. Wartime motivates bureaucratic flexibility that will be potentially hard to preserve through institutionalization. War has also supplied a difficult-to-replace, without legislative cooperation, immediate flexible source of funds—supplemental funding. If future rapid acquisitions have to compete with other programs in the standard POM cycle, delivery of the capabilities will cease to be rapid in the same way as currently.

War also motivated both the Army and DoD as a whole to establish structures dedicated to rapid acquisitions serving the needs of the conflicts. This wide variety of processes and organizations has supplied a large number of capabilities for DoD and the Army. However, these designated processes have not met every rapid acquisition need of the Army, and it may not be feasible to expect them to do so. For instance, the programs we examined in the case studies did not rely heavily on these named processes.

One finding that was surprising to our research team was the degree to which successful rapid acquisitions of C2 leveraged other existing related programs to conduct the acquisition. Using existing requirements, contracts, and acquisition documentation from other programs speeded the acquisition of both JNN-N and CPOF. In particular, JNN-N used Engineering Change Proposals on existing contracts for very roughly similar equipment to initiate and continue the purchases, and sole-source contracts to speed the acquisition considerably. As a different example, CPOF inserted itself into (and later "swallowed") a related program of record, one of whose requirements CPOF satisfied. The Army eventually inserted both JNN-N and CPOF into other programs of record to satisfy long-term sustainment and oversight requirements.

To make arrangements for sustainment, previous rapid acquisitions of C2 have provided initial support via multiyear contractor warranties, which provide the Army with the option of either extending the arrangement or later substituting properly trained Army personnel. In the cases we considered, we found that rapid acquisitions do not consider lifecycle issues sufficiently to satisfy the long-term interests of the institution. To sustain the capability in the long term, it has been necessary to transition the rapid acquisition into a standard acquisition via either an established or ad hoc process. Regardless of the method of the transition, we found that careful transition planning and staffing consistency for both the program office and the contractor is central to the outcome.

Even after many years, very little is actually known about the efficiency or effectiveness of these urgent need and rapid acquisition activities, and there has been no systematic attempt to capture best practices. The Army has not collected data that would provide insight into how well urgent needs are met, how well rapidly acquired systems perform, or how well these processes are working. There may have been follow-on efforts to track deployment, provide fielding support, and obtain user feedback for specific systems (as has been the case for JNN and CPOF), but to our knowledge this information has not been accumulated and analyzed with the intent to improve rapid acquisition processes generally. Interestingly, the urgent needs, rapid acquisition, and especially the transition processes (CDRT) have not been well known, and training that deployed forces have received in the use of urgent needs and rapid acquisition processes has been minimal.

Institutionalizing rapid acquisition capabilities presents challenges in at least two forms. First, replicating those aspects of the wartime environment that enable rapid acquisition—including separate funding, a sense of urgency, and the willingness to accept increased operational risk and less than 100 percent of desired performance—will be more difficult in peacetime. Second, accommodating the legitimate concerns of certain stakeholders whose subject areas or functions were downplayed to some extent to enable rapid acquisition will slow the acquisition.

On the other hand, institutionalization provides other opportunities for the Army to improve its ability to conduct rapid acquisition. The Army can explore ways to institutionalize organizational tolerance of increased risk, and to contain the sway of traditional acquisition system stakeholders. Institutionalizing rapid acquisition could also establish mechanisms that improve a program manager's abilities to tailor their acquisition strategies. It would also likely reduce the time staff needs to spend in inventing ad hoc methods to perform rapid acquisition, and reduce the learning curve for staff seeking to perform rapid acquisition for the first time. As a transition process, CDRT could be both strengthened and publicized, then incorporated into the institutionalized process.

While we ourselves see war as a key driver for successful rapid acquisition, and question how it may continue to be "rapid" in the long term without a war to motivate the necessary bureaucratic compromises, there appears to be a conclusion within DoD that institutionalizing the continuing urgent need and rapid acquisition experience is desirable. However, there is an ongoing debate about how best to do this. One aspect of that debate is whether institutionalization requires the creation of a new organization responsible for implementing rapid acquisition processes, or whether the rapid acquisition processes can be created and formalized, but then implemented within the existing structure of the Army (and DoD) acquisition organizational structure.[42] The Army does have a possible mechanism in place that could accommodate both rapid acquisition of C2 systems and their IT-unique attributes: capability packages are aligned with the Army's two-year cycle force generation model, unit set fielding.[43]

---

[42] See, for instance, Fawzia Sheikh, "DoD Dismisses Call for New Acquisition Agency to Meet Urgent Needs," *Inside the Pentagon*, April 15, 2010.

[43] See Kate Brannen, "Dempsey: Rapid Pace of Change Means Shorter Development Time Lines," *Inside the Army,* Vol. 21, No. 51, December 28, 2009.

Figure 4.2. Summary of Recommendations

Summary of Recommendations

- **Regularly & systematically capture "Lessons Learned" from rapid acquisitions**

- **Establish and support flexible mechanisms for funding acquisition efforts outside of the 2-yr. POM cycle, without supplemental funding.**

- **Promote awareness of and strengthen existing Army processes for transition of non-PORs to official standing, such as CDRT**

- **Establish ways to expedite testing to support rapid acquisition**

- **Train the institution to expect PMs to tailor their acquisition strategy and mechanisms used**

- **Enable PMs to prioritize and make trade-offs in meeting requirements based on user feedback**

- **View acquisition documents, staff, and contracts of existing programs as potential enablers of rapid acquisition**

- **Require PMs to assess planned rapid acquisitions for inclusion of the enabling factors discussed here**

RAND   Army Rapid Acquisition of C2   **26**

There are a number of measures the Army can take to support rapid acquisition. These recommendations include establishing flexible funding sources in the absence of supplemental funding and promoting awareness of existing transition processes. We also recommend that the Army make improvements in testing considerations, documentation of lessons learned (a recommendation discussed in conjunction with the next slide), and changing institutional expectations for program managers. Finally, based on the experiences of programs examined in this work, it is clear the Army should view related existing programs as resources to support rapid acquisitions.

To support rapid acquisition in the long term, the Army should find a way, with the support of Congress, to establish a dedicated urgent needs funding line in the Army budget to provide an immediately taskable source of funds for new rapid acquisitions. Of course, Congress would need to approve such a budget line item and would also have to grant the Army additional budget flexibility within that line item to respond to changing needs. This recommendation matches a recent recommendation from the National Research Council (NRC) regarding acquisition of IT systems that DoD "work with Congress to explore how to make use of flexibility consistent with current legal requirements. The NRC notes that "acquisition funds are sometimes allocated by

Congress to a larger mission or program area or in some cases to a portfolio of projects identified with an area of mission need . . ." In addition, the NRC recommends that Congress allocate funding to mission areas in the longer term. It states that DoD uses this type of funding for maintenance upgrades to aircraft avionics software.[44] Fast and flexible funding is essential to rapid acquisition, and it is difficult to imagine any such acquisitions occurring without it. Perhaps the Army should "register" concepts when they are conceived, and then track for each how long it takes them to receive initial funding, as a metric for the rapidness of the start. This metric would enable the Army to present data to Congress on the effects of having a flexible source of funding.

The Army should publicize the existence of the process called CDRT that can help program managers who have delivered a useful capability to the warfighter with to sustain their capability. The Army can choose to include the capability in their long-term inventory, or not, through this process. If the Army chooses through CDRT that the capability will either temporarily "sustain" or have the capability to "endure," the institutional Army will support the necessary support arrangements.

The Army should also consider ways to expedite testing to support rapid acquisition. The testing community could be asked to develop proposals on how to expedite testing for rapid acquisitions and, equally importantly, how to communicate the resulting risk tradeoffs the Army makes by doing so. For systems that have been used operationally, the testing community should seek to establish mechanisms to "give credit" during testing for operational use.

It would also be supportive to the conduct of rapid acquisition for the Army to train personnel and oversight managers to expect program managers to tailor their acquisition strategy and the mechanisms they use to perform the acquisition. The *Defense Acquisition Guidebook* discusses the importance of tailoring a program's acquisition strategy to fit the program,[45] but in reality, it has seemed anecdotally hard for program managers to omit any step, stakeholder, or document in performing a rapid acquisition.

Beyond the tailoring of the Acquisition Strategy, the Army should encourage and protect program managers when they seek support to make tradeoffs to support rapid delivery of a needed capability. If decision-making authority over funding for requirements is held at a very high level, in effect the needed tradeoffs cannot be made efficiently by those who know the most about the system and the problems sets it addresses. These tradeoffs should be informed and both requirements and the degree of their satisfaction prioritized during development using concurrent user feedback.

---

[44] National Research Council, *Achieving Effective Acquisition of Information Technology in the Department of Defense*, Washington, D.C.: The National Academies Press, 2010, p. 8.

[45] *Defense Acquisition Guidebook*, "Tailoring," Section 2.2.1.2, February 19, 2010. The most current DAG is available online at https://dag.dau.mil/.

Army program managers and oversight authorities should take note of any related acquisition programs before or during the course of a rapid acquisition, as these programs, rather than being competition, are likely enablers of the rapid acquisition if they make a timely demonstration of a useful capability. Program managers should consider existing requirements, acquisition documents, contracts, and program staff as potential resources, given the necessary institutional support.

In addition, the Army should require program managers to assess any acquisitions they plan to carry out in a rapid manner for the presence or absence of the key enabling factors identified in Figures 2.12 and 2.13, as the programs we evaluated in this work that encompassed all or most of these factors were able to conduct a rapid acquisition successfully.

Figure 4.3. Document and Preserve Recent Rapid Acquisition and Transition Experience

> # *Document and Preserve*
> # *Recent Rapid Acquisition and Transition Experience*
>
> - **Additional case study and multi-case analyses to document lessons**
>
> - **Assess possible process duplication**
>
> - **Develop and implement metrics to provide insight into efficiency and effectiveness**
>
> - **Make criteria for transition decisions explicit**
>
> - **Maintain up to date set of relevant transition enabling mechanisms**
>
> **RAND**   Army C2 **final brief 32** 2 January 2013

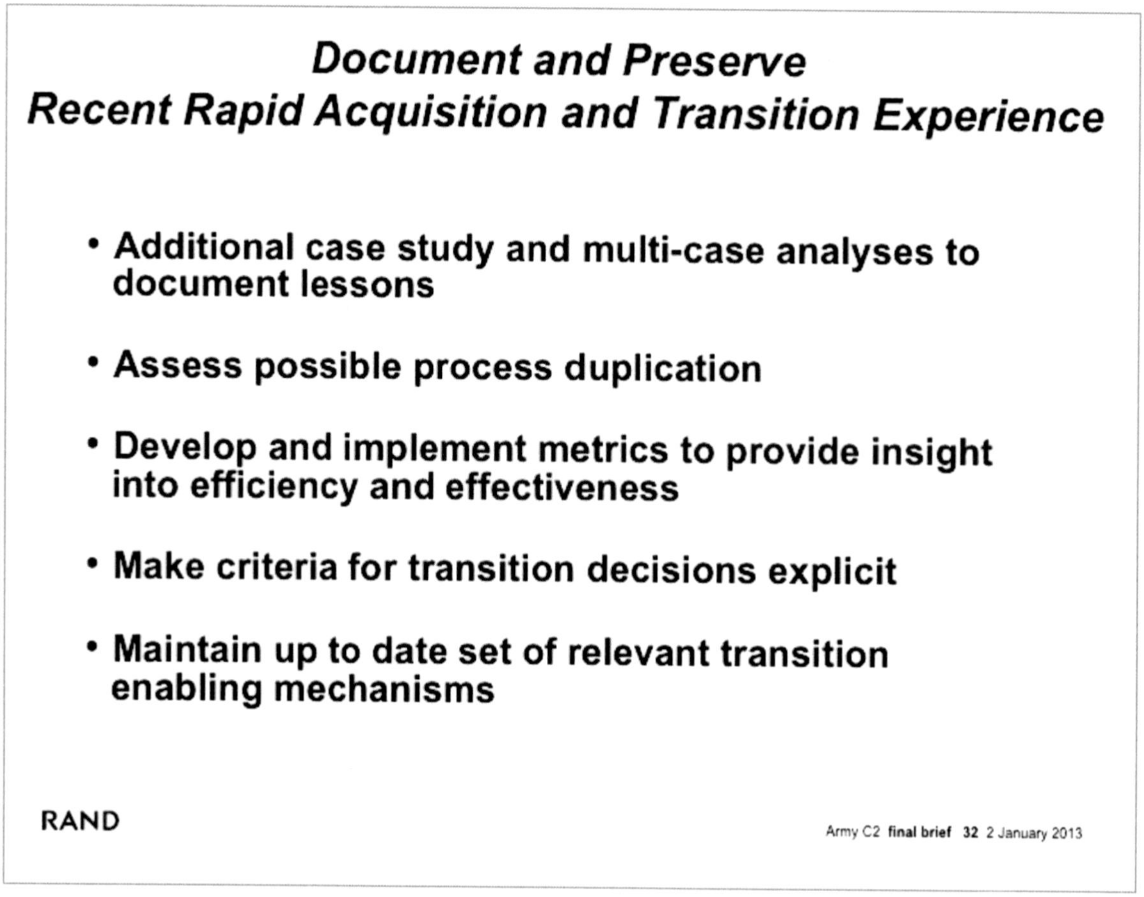

It is important that the Army work now to document its recent experiences with rapid acquisition. Future attempts to perform rapid acquisition will have to "start over" in developing methods and best practices if the Army does not preserve its current knowledge base in this area. While our work has documented three case studies, the Army could undoubtedly capture even more useful lessons learned and best practices by documenting lessons from a wide array of rapid acquisitions, both C2 and otherwise.

Currently there is little empirical evidence on how effective and efficient officially established rapid acquisition processes have been. As a result, we are recommending that DoD and the Army take the time now to capture lessons from recent experience that would enable a new permanent rapid acquisition process to address the challenges and take advantage of the opportunities discussed previously. This includes performing additional case studies of rapidly acquired C2 and other weapon systems to validate the results to date. The Army should also perform a more thorough review of existing processes to better understand how they work and to identify duplication and coordination issues. As a matter of standard practice, the Army should develop and implement a system to collect information on rapid acquisition programs and processes to provide some insight into how well and how efficiently they have satisfied warfighter needs.

Capturing lessons from rapid acquisitions systematically will enable the Army to propose and vet metrics for efficiency and effectiveness in rapid acquisition. If sufficiently robust, the efficiency metric could then potentially be extended to measure the rapidity of any type of acquisition. An effectiveness metric might also enable the Army to better characterize the degree of performance compromises made with any rapid acquisition.

Institutionalized study of rapid acquisitions could also help with transitioning rapid acquisitions to the mainstream for sustainment. For instance, a careful study could make the criteria for transition decisions and timelines explicit rather than ad hoc, if appropriate. Capturing lessons learned would also enable the Army to maintain a portfolio of available transition options for managers of future acquisitions.

It is important to recognize that the Army's traditional acquisition community (PEOs, program managers, ASA(ALT)) has supported many successful rapid acquisition efforts, including C2. These efforts, more broadly, have included the MRAP program, Stryker fielding and add-on armor, as well as the JNN-N program we have discussed in this report. Significantly, rapid acquisitions have not necessarily required formal dedicated "rapid" organizational processes or structures. The implication here is that rapid acquisition can be done both with and without relying on the traditional acquisition process and organizations or, more likely, some combination of within and outside that process. The sense of urgency conveyed by wartime, as well as the program office's creative use of available processes, authorities, tailoring, contracting mechanisms, COTS/GOTS, and flexible requirements, among others, appears to be the real difference between "rapid" acquisition of C2 and more traditional acquisition.

(This page is intentionally blank)

Joint/Army Rapid Capabilities and Materiel Developments Initiatives

To understand how rapidly acquired C2 systems can be better managed, we identified the Army and DoD (joint) rapid capabilities and developments initiatives that currently exist. This includes processes that identify and prioritize, fund, procure, and transition or terminate non-POR systems that resulted from urgent needs requests from operational commanders. Identifying current processes is difficult because there are a number of processes across DoD at both the service and OSD levels. In a July 2009 study, the Defense Science Board (DSB) found that

> over 20 different ad hoc organizations within the Joint Staff, the Office of the Secretary of Defense, and each Service now have urgent needs processes. The procedures these organizations have developed to generate, validate, and fulfill warfighting requirements vary across the DoD.[46]

Given the large number of processes, this appendix will only focus on Joint/Army processes.

In May 2010, the Army Force Management School (AFMS) released version 15 of its "Capabilities Development and System Acquisition Management: 2010 Executive Primer." Within this primer, the AFMS identified several Joint/Army rapid capabilities and materiel developments processes "that provide timely support to Soldiers deployed in combat, while facilitating Army transformation."[47] This appendix provides a brief description of a select number of processes that were found in the DSB report and the AFMS primer. It will also briefly look at which processes work together. Finally, we review the challenges that these processes face, based on available information.

Table A.1 provides the names of the processes and their primary function: whether they identify, prioritize, fund, procure, or transition non-POR systems resulting from urgent needs requests. Most of these organizations and processes are single function. For instance, one process validates the need, another identifies the solution and acquires it, a third provides the funding, and a fourth decides whether to transition the system to a POR. Some of these single function processes are closely related; the Army's ONS, AR2B, and REF processes generally support each other and are roughly equivalent to more traditional requirements, budgeting, and acquisition processes, respectively. These

---

[46] Defense Science Board (DSB), *Fulfillment of Urgent Operational Needs*, July 2009, p. 9.

[47] Bob Keenan, "Capabilities Development and System Acquisition Management: 2010 Executive Primer," Army Force Management School (AFMS), Version 15, May 2010, p. 89.

processes also seem to go through a similar pattern as peacetime processes: capability gap/need is identified by warfighter; then validated; then solutions are sought, funded, and contracted. The key distinction between peacetime and wartime processes is that the wartime processes are much faster and more focused than peacetime processes.

Table A.1
Select Joint/Army Rapid Capabilities and Materiel Developments
Initiatives, Processes, and Organizations

Name of Process	Joint/ Army	Purpose
Operational Needs Statement (ONS)	Army	Identify UONs
Immediate Warfighter Need (IWN)	Joint	Identify UONs
Joint Urgent Operational Need (JUON)	Joint	Identify UONs
Army Requirements and Resourcing Board (AR2B)	Army	Identify and Prioritize UONs
Joint Rapid Acquisition Cell (JRAC)	Joint	Identify and Prioritize UONs
Quick Reaction Fund (QRF)	Joint	Fund UONs
Rapid Reaction Fund (RRF)	Joint	Fund UONs
Rapid Equipping Force (REF)	Army	Identify and Procure UONs
Rapid Fielding Initiative (RFI)	Army	Procure UONs
Joint Improvised Explosive Devices Defeat Organization (JIEDDO)	Joint	Procure UONs
Mine Resistant Ambush Protected (vehicle) Task Force (MRAP TF)	Joint	Procure UONs
Rapid Reaction Technology Office (RRTO)	Joint	Procure UONs
Capabilities Development for Rapid Transition (CDRT)	Army	Transition UONs
Immediate Warfighter Needs (IWN) Process	Army	Transition UONs

The first five processes in Table A.1 identify urgent needs in the field through the collection of key operational need information. In addition to identifying needs, two of the processes—Army Requirements and Resourcing Board (AR2B) and Joint Rapid Acquisition Cell (JRAC)—also prioritize urgent needs for the Army and DoD.

Operational Needs Statement (ONS)

The Operational Needs Statement (ONS) is the main way for the Army field commanders to identify and document urgent operational needs. This process began in 1987 but was not heavily utilized until Operation Enduring Freedom (OEF) and Operation Iraqi Freedom (OIF). To illustrate how frequently this process is currently being used, the GAO gathered the following statistical information on the ONS process: "From September 2006 to February 2010 the Army's database shows 6,712 Operational

Needs Statements containing 21,864 urgent needs requests that have been or are being processed to support operations in those two theaters [Iraq and Afghanistan]."[48]

The response time for fulfilling an ONS varies based on the criticality of the need. It is calculated based on two parts: the first is ONS validation and staffing, and the second is developing, testing, and producing a solution. The earliest that a materiel solution can be fielded is 30 days based on two criteria: the unit is preparing or is already deployed, and a solution already exists.[49] Typically, if COTS exists, then the process takes three to six months, while new technology takes approximately 12 to 18 months. Otherwise, the ONS process can take up to 120 days. The process has been used to fulfill a variety of needs, including "new capabilities to shortfalls of existing equipment in theater, to requests for training equipment for mobilizing units in the United States."[50] After an ONS is identified, the Army Requirements and Resourcing Board certifies the need. In addition, this process is overseen by the Office of the Deputy Chief of Staff G-3/5/7.

Immediate Warfighter Need (IWN)

Created in 2004, an Immediate Warfighter Need (IWN) is similar to the Army ONS, but it satisfies joint needs. IWNs are a portion of the larger set of joint urgent operational needs (JUONs) where a 120-day or less solution is critical; however, it had been difficult to find and field a solution within 120 days, so the maximum deadline has been extended to two years.[51] The Joint Rapid Acquisition Cell (JRAC) reviews JUONs and designates IWNs from the larger set. The JRAC is also responsible for evaluating the resourcing for the IWNs. Given the critical deadline, it is common for IWNs to be highly visible at the Office of the Secretary of Defense and the Deputy Secretary of Defense levels.[52]

---

[48] GAO, April 2010, p. 9.

[49] COL(P) Peter N. Fuller, "Rapid Acquisition — Developing Processes That Deliver Soldier Materiel Solutions Now," United States Army Acquisition Support Center, February 2008.

[50] GAO, April 2010, p. 9.

[51] GAO, April 2010, p. 17.

[52] Norton A. Schwartz, (Lieutenant General, USAF Director, Joint Staff), "Chairman of the Joint Chiefs of Staff Instruction: Rapid Validation and Resourcing of Joint Urgent Operational Needs (JUONs) in the Year of Execution (CJCSI 3470.01)," Joint Chiefs of Staff, July 15, 2005 (current as of July 9, 2007), p. GL-1.

Joint Urgent Operational Need (JUON)

In the Department of Defense, JUONs are urgent needs, identified by a combatant commander, that span more than one service.[53] The JUON process was started in November 2004 based on a large increase in operational needs in Iraq and Afghanistan. JUONs are limited to addressing needs that

> fall outside of the established Service processes . . . and . . . if not addressed immediately, will seriously endanger personnel or pose a major threat to ongoing operations.[54]

These joint needs tend to be more difficult and costly to address than needs that only deal with one particular service, so JUONs required more signatures for approval than ONS.

JUONs are initially identified and validated by the Joint Staff and JRAC. Then they are sent through a process like the JIEDDO or the services' rapid acquisition processes to be fulfilled in a rapid manner. JUONs take anywhere from three to six months to satisfy needs using COTS and 12 to 18 months for new technologies. From 2005 to May 2009, there were approximately 228 evaluated JUONs.[55]

Army Requirements and Resourcing Board (AR2B)

The Army Requirements and Resourcing Board was established in December 2004. Its predecessors were the Army Strategic Planning Board (ASPB) and the Setting the Force Task Force. Former Vice Chief of Staff, retired General Richard Cody, came up with AR2B during 2002 out of frustration with the slow prioritization of resources in the Army.

> The AR2B is the mechanism (forum) for validating, prioritizing, and resourcing critical operational needs (ONS and ESDs) for rapid senior leadership decision-making (accelerated fielding solutions) in support of a named operation. The AR2B identifies solutions in the year of execution and/or budget year that require possible resource realignment.[56]

---

[53] According to a recent GAO report (GAO, April 2010), the terms "urgent operational need" and "immediate warfighter need" used to have slightly different meanings, but now, UON and IWN have been subsumed into the term "JUON" in practice.

[54] DSB, July 2009, p. 10.

[55] DSB, July 2009, p. 22.

[56] Bob Keenan, "Capabilities Development and System Acquisition Management: 2010 Executive Primer," Army Force Management School (AFMS), Version 15, May 2010, pp. 90–91.

From 2004 to the fourth quarter of fiscal year 2008, the AR2B processed over 8,900 requests for equipment.[57]

There are three parties that make up the AR2B: the Deputy Chief Of Staff, G-3/5/7; the Deputy Chief of Staff, G-8; and the Military Deputy, Assistant Secretary of the Army (Financial Management and Comptroller) (ASA(FM&C)). Along with the above parties, ONS are reviewed by Army Staff Offices and Field Commands. The ONS are then validated and prioritized by the AR2B. The step of prioritization is what differentiates the AR2B from ONS and JUONs. The board also tries to come up with the funding for each capability. Each of the above functions is expedited in order to get the capability to the field rapidly.

Joint Rapid Acquisition Cell (JRAC)

The JRAC, which started in September 2004, is a joint organization in OSD that

> facilitate[s] meeting the joint urgent operational needs (JUONs) of the Combatant Commanders as validated by the Joint Staff; and serve[s] as the point of contact on the OSD staff for tracking JUON resolution."[58]

After JUONs/IWNs are identified, JRAC assigns the JUONs/IWNs to the appropriate service/joint organization for procurement. The JRAC then tracks and monitors the JUONs until they are fulfilled. Critical decisions are attempted by the JRAC within 2 to 14 days; however, more complicated needs could take up to two years to fulfill. Since 2004, the JRAC has received an estimated 225 joint urgent needs requests.[59]

JRAC involves many different representatives throughout DoD. The core member group consists of the following parties: USD(AT&L), USD(Comptroller), the DoD General Council, and the Joint Staff. An advisory group consists of USD(I), USD(P&R), USD(P), ASD(NII), Director, Program Analysis and Evaluation (PA&E), combatant commanders, military services, and the Director, Operational Test and Evaluation.

> The advisory group supports the core group based on the specific immediate warfighter need request and functions in a manner similar to an overarching integrated product team.[60]

---

[57] CPT John H. Dabolt IV, "Army Requirements and Resourcing Board Rapid Reaction in an Era of Persistent Conflict," *The Oracle*, Vol. 4, 4th Quarter, FY 2008, p. 5.

[58] "Rapid Fielding Directorate: Joint Rapid Acquisition Cell," Joint Rapid Acquisition Cell website, June 2010.

[59] GAO, April 2010, p. 31.

[60] Schwartz, p. GL-1.

In the past couple of years, the JRAC has gone through some restructuring at the OSD level.

> Until 2008, OSD directed that the JRAC report to the Secretary of Defense, through the Under Secretary of Defense for Acquisition, Technology and Logistics (AT&L) and the Under Secretary of Defense (Comptroller), for monitoring and tracking joint urgent needs, facilitating the identification and resolution of issues, and providing regular status reports to the Secretary and Deputy Secretary of Defense. Since 2008, the Under Secretary realigned the JRAC within the Office of the Director for Defense Research and Engineering (DDR&E).[61]

In order to consolidate several of the joint rapid acquisition processes within DoD, a Director for Rapid Fielding was appointed in September 2009. As of this writing, the Director for Rapid Fielding oversaw the following three offices: the Rapid Reaction Technology Office, the Complex Systems Office, and the Joint Rapid Acquisition Cell.[62] The Director of Rapid Fielding reported to the Director, Defense Research and Engineering (DDR&E). The DDR&E reported to the Under Secretary of Defense for Acquisition, Technology, and Logistics (USD(AT&L)) and the Secretary and Deputy Secretary of Defense.

The figure below provides a look at the Joint Chiefs of Staff and Joint Rapid Acquisition Cell Process for reviewing, validating, and fulfilling JUONs:

---

[61] GAO, April 2010, pp. 1–2.

[62] "Rapid Fielding Directorate: About RFD," Deputy Assistant Secretary of Defense, Rapid Fielding website, June 2010.

Figure A.1. Joint Chiefs of Staff and Joint Rapid Acquisition Cell Process for Reviewing, Validating, and Fulfilling JUONs

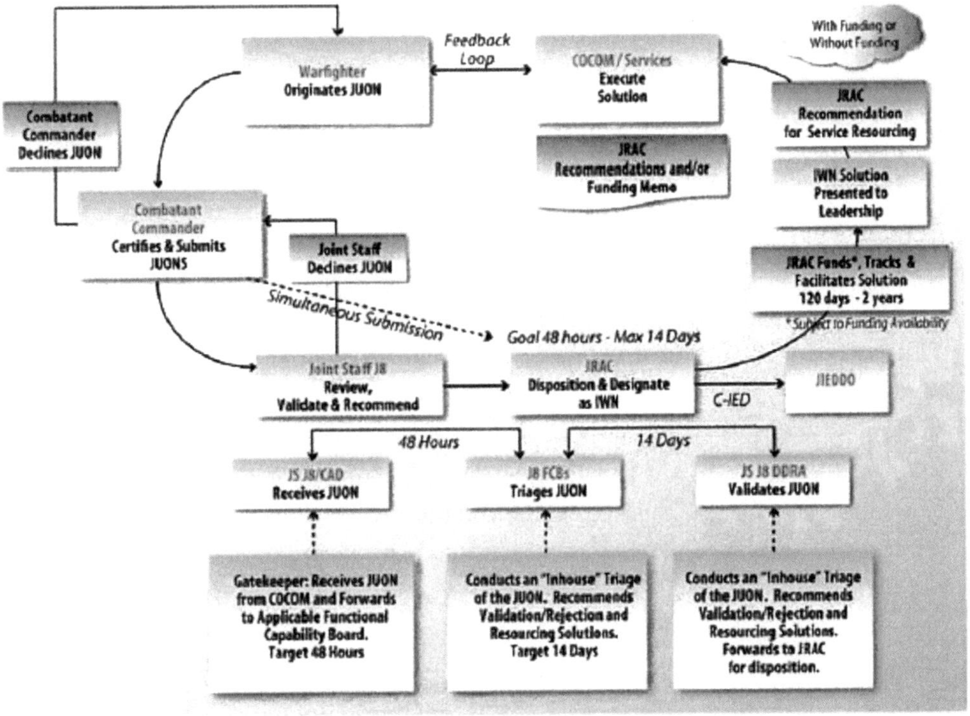

SOURCE: Defense Science Board, July 2009, p. 43; Joint Rapid Acquisition Cell, February 19, 2009.

Quick Reaction Fund (QRF)

The Quick Reaction Fund (QRF) is one of two joint processes discussed in this appendix that are sources of funding for rapid capabilities. QRF was established by Congress in fiscal year 2003. The focus of the QRF is "responding to emergent needs during the execution years that take advantage of breakthroughs in rapidly evolving technologies."[63] The projects tend to finish in less than one year. The QRF also supports research that fills gaps in DoD acquisition programs. Finally, it fulfills critical operational needs using mature technologies. The QRF is managed by the Director of Defense Research and Engineering (Plans and Programs).

---

[63] Glenn Fogg, "How to Better Support the Need for Quick Reaction Capabilities in an Irregular Warfare Environment: Quick Reaction and Rapid Quick Reaction Funds," Rapid Reaction Technology Office (RRTO), April 21, 2009, p. 4.

Rapid Reaction Fund (RRF)

The Rapid Reaction Fund (RRF) provides funding to projects that focus on counterterrorism or counterinsurgency. It is a joint fund that, like QRF, is managed by the Rapid Reaction Technology Office (RRTO). RRTO and JRAC are both directly under the Director of Rapid Fielding in the office of the Director, Defense Research and Engineering. RRF supports spiral development on projects ranging generally from six to 18 months. Given that RRF is a joint fund, it is able to draw on technology advancements and programs throughout DoD.

The next five processes are ones that rapidly procure capabilities. The Rapid Equipping Force and Rapid Fielding Initiative are both Army processes, while JIEDDO, the MRAP Task Force, and the Rapid Reaction Technology Office are joint offices.

Rapid Equipping Force (REF)

The Rapid Equipping Force (REF) began in 2002, and is an organization dedicated to rapidly resolving urgent warfighter needs for which no solution currently exists in the Army inventory. REF now resides within Headquarters Army G-3/5/7. REF is responsible for providing Army units deploying and predeploying globally with rapidly acquired solutions to satisfy urgent operational needs.[64] The solutions are tailored to the specific needs of units, and limited quantities are typically produced. REF tries to provide critical solutions within 90 days and others within 180–360 days. The critical solutions are typically filled through integrating COTS and GOTS equipment in a manner geared to specific capability shortfalls. REF identifies other solutions by searching the public and private sector for emerging technologies.

REF relies on a shortened requirements process, a flexible source of funding, and direct access to senior Army decision makers to expedite its acquisitions. In many cases REF uses a "10-liner" submitted by units identifying capability gaps in order to identify UONs. In some cases the UONs are also formalized as ONS;[65] however, in others, the only statement of the need is the 10-liner.[66] REF prioritizes the use of its resources to fill

---

[64] Headquarters, Department of the Army (HQDA), Deputy Chief of Staff G-3/5/7, REF Annex to the 2013 Army Posture Statement, December 2012.

[65] An "ONS" is an Operational Need Statement that identifies "an urgent need for a non-standard and or [sic] unprogrammed capability to correct a deficiency or improve a capability to enhance mission accomplishment." Office of the Deputy Chief Management Officer Representative, "Review of Acquisition Processes for Rapid Fielding of Capabilities in Response to Urgent Operational Needs" (draft version), Department of Defense, October 21, 2012.

[66] Office of the Deputy Chief Management Officer Representative, "Review of Acquisition Processes for Rapid Fielding of Capabilities in Response to Urgent Operational Needs" (draft version), Department of Defense, October 21, 2012.

various identified capability gaps using what it calls a "REF Integrated Priority List (RIPL)."[67] Currently the RIPL notes that two-thirds of the requirements REF is working to satisfy focus on the following areas:[68]

- Dismounted IED defeat
- Dismounted operations support
- Intelligence, surveillance, and reconnaissance shortfalls in environmentally inhospitable operating environments
- Small combat outposts and patrol bases force protection and sustainment

REF finds its solutions for the first of these areas, improvised explosive devices (IEDs), in coordination with JIEDDO.[69]

Since 2005, REF has supplied 1,979 different types of equipment to deployed units and Combat Training Centers.[70] Examples of types of equipment REF has fielded recently include: "Dismounted IED Defeat (Minotaur)," tethered tactical aerostats, improved gunner restraints, and an integrated blast effects sensor suite.[71] REF does not supply items that can be procured through the traditional acquisition process or items that can be supplied through the military supply chain, nor does it field solutions to the entire Army, as is the case with the Rapid Fielding Initiative.

Figure A.2 provides a look at REF's decision-making process. Proposed REF projects can enter the REF process in one of three different ways: either REF itself identifies an emerging need for which a solution exists, REF obtains technology via external suggestion or its own searches that it deems likely to be useful to the warfighter, or the warfighter submits a 10-liner statement of need to REF. If the REF director decides it is appropriate to pursue the project, REF assesses whether it is possible to "harvest" the necessary solution from existing military equipment, or whether it should procure its production from an outside vendor. After development of a Theater Deployment Plan, REF or the vendor ships the solution to the theater and gathers operator feedback on its utility. The Army then must decide whether to field the capability more broadly to the rest of the Army. Regardless of whether the capability is deployed broadly or only to a few personnel, while the equipment is in the hands of the warfighter, logistics personnel must plan to sustain it.

---

[67] Edward Jozwiak, in-person conversation with Shara Williams, February 7, 2013.

[68] Rapid Equipping Force, "REF Integrated Priority List," no date.

[69] Edward Jozwiak, in-person conversation with Shara Williams, February 7, 2013.

[70] Edward Jozwiak, "Rapid Equipping Force (REF) Info: Q. Request for information on REF (UNCLASSIFIED)," email to Shara Williams, February 25, 2013.

[71] Headquarters, Department of the Army (HQDA), Deputy Chief of Staff, G-3/5/7, REF Annex to the 2013 Army Posture Statement, December 2012.

Figure A.2. Rapid Equipping Force's Decision-Making Process

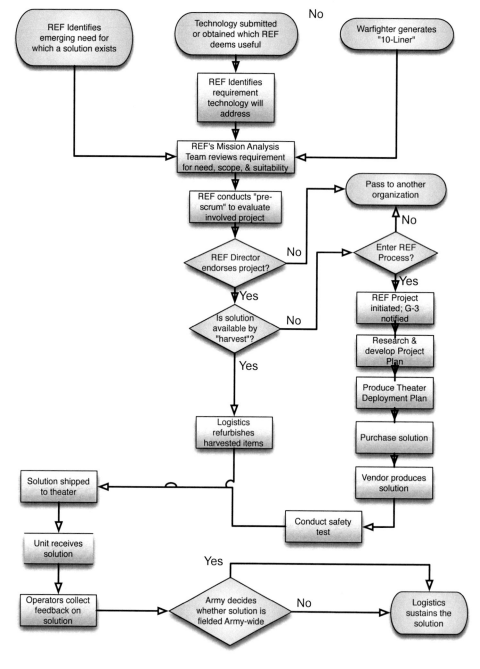

SOURCE: Office of the Deputy Chief Management Officer Representative, "Review of Acquisition Processes for Rapid Fielding of Capabilities in Response to Urgent Operational Needs" (draft version), Department of Defense, October 21, 2012.

Rapid Fielding Initiative

The Rapid Fielding Initiative was created in 2002 in response to the demands of OEF and OIF. RFI's intent is to "provide capabilities procured and distributed through an organized program vice ad hoc purchasing by units and Soldiers."[72] It is under the Army' Program Executive Office (PEO) Soldier.[73] The office's mission is to design, develop, procure, and field standard equipment to soldiers. The RFI list includes two types of equipment: equipment every soldier receives (such as helmet, clothing items, etc.), and equipment fielded to units. This is different from the Rapid Equipping Force because REF provides specific solutions for units rather than standard equipment intended to be distributed Army-wide. RFI was created because PEO Soldier saw the need for streamlining the process for distributing equipment to deploying units. Since its creation, RFI has also found that duplicative fieldings are also a problem. The office now strives to implement something called lean fielding in order to reduce duplication. As of May 2010, RFI has said that it equipped nearly 1.5 million deploying soldiers.

Over the next 12 months, PEO Soldier will be going through a restructuring in order to integrate RFI into a G-4 directorate. This is being done

> in order to institutionalize the benefits of RFI across the PEO Soldier portfolio. In addition, several of the project manager positions will be refocused to a smaller range of products. While no products will be removed from PEO Soldier's portfolio, these changes will allow for a faster, more focused response.[74]

Joint Improvised Explosive Devices Defeat Organization (JIEDDO)

The Joint Improvised Explosive Devices Defeat Organization (JIEDDO) began in October 2003 as an Army IED Task Force. At that time, IEDs were being used in Iraq and Afghanistan at an alarming, rate, which led to the formation of the task force by the Army Chief of Staff. The Army used a variety of sources to come up with technology that would protect soldiers using a steady amount of rapid acquisition funding. However, the IED threat was DoD-wide, so the task force became a joint task force in 2004. Then in 2005, the task force gained even greater importance when it began reporting directly to the Deputy Secretary of Defense. As a joint task force, funding and authority increased. In February 2006, the joint task force became a permanently manned entity called the Joint IED Defeat Organization (JIEDDO).

---

[72] Colonel Thomas H. Bryant, "Rapid Fielding Initiative Overview to the 2005 Acquisition Senior Leaders and AMC Commanders Conference," *Rapid Fielding Initiative*, August 23, 2005, p. 7.

[73] PEO Soldier is only responsible for what a soldier "wears or carries."

[74] "Individual Equipment and Weapons," Association of the United States Army, Army Magazine, October 2009, p. 376.

JIEDDO and MRAP have created some important solutions that are protecting soldiers from deadly threats in both Iraq and Afghanistan. Also, they are receiving a significant amount of funding. Of the $50 billion in available rapid acquisition funding from 2005 through May 2009, JIEDDO received 15.7 percent or $7.85 billion and MRAP received 24 percent or approximately $12.5 billion.[75] This relatively large amount of rapid acquisition funding has resulted in these organizations having more exposure to Congress' scrutiny and to auditing agencies like the GAO, which has not been as common among the other smaller urgent needs and rapid acquisition organizations.

Figure A.3 is a diagram of the various pieces of the JIEDDO capabilities and acquisition process.

Figure A.3. JIEDDO's Capabilities and Acquisition Process

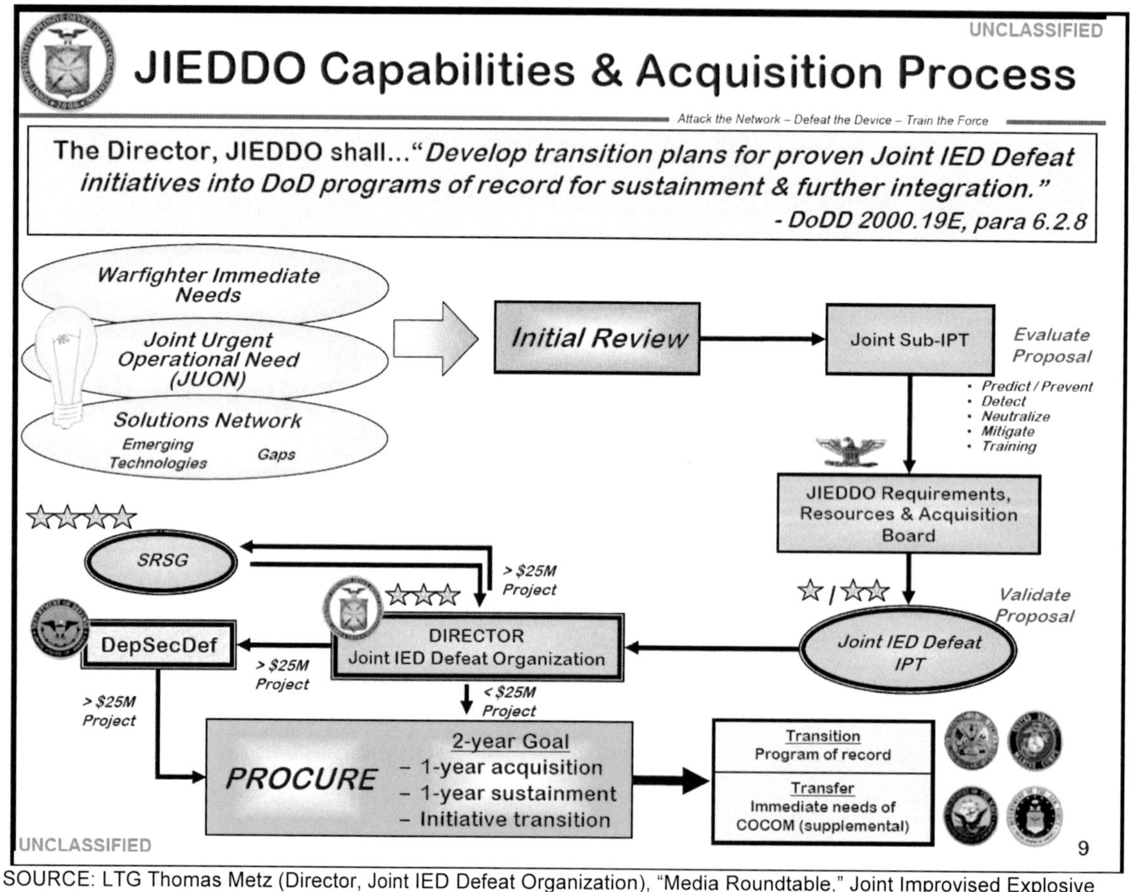

SOURCE: LTG Thomas Metz (Director, Joint IED Defeat Organization), "Media Roundtable," Joint Improvised Explosive Device Defeat Organization, April 30, 2008.

---

[75] DSB, July 2009, p. 11.

Mine Resistant Ambush Protected (vehicle) Task Force (MRAP TF)

DoD initiated the MRAP program in February 2007. Two months later, Secretary of Defense Gates said that MRAP was DoD's most important acquisition program given the number of U.S. soldiers killed by IEDs in both Iraq and Afghanistan. Secretary Gates gave it a DX rating,[76] which allowed the MRAP task force significant access to materials and funds. Such open access is unusual for less critical acquisition efforts. MRAPs received priority funding, and the Marine Corps was placed in charge of procuring the vehicles. DoD used a tailored acquisition approach with flexible requirements and an inflexible schedule to rapidly procure MRAPs, thus forcing the task force to utilize existing technologies.

The need was first identified in February 2005 and initial operational capability was in October 2007. The program has been viewed as highly successful because it provided a critical operational solution.

> In July 2008, nearly all testing was completed; the Marine Corps had placed orders for 14,173 MRAPs; and, as of May 2008, 9,121 vehicles had been delivered. As of July 2009, 16,204 vehicles have been produced and 13,848 vehicles fielded in two theaters of operation.[77]

As of October 2009, GAO said that "about $22.7 billion has been appropriated for the procurement of more than 16,000 MRAP vehicles."[78]

The MRAP program still has some challenges that need to be addressed, like the creation of a long-term sustainment plan for the vehicles being used in Iraq and Afghanistan. Also, DoD has to decide how MRAPs will fit into any future DoD organizational plans. There also needs to be some modification of the designs for MRAPs to be effective in both Iraq and Afghanistan.

Rapid Reaction Technology Office (RRTO)

The Rapid Reaction Technology Office (RRTO) is a joint office, formerly known as the Combating Terrorism Technology Task Force (CTTTF). CTTTF began in 2003, and

> was stood up to provide rapid response to operations in Iraq and other theaters in support of the Global War on Terrorism (GWOT) and to accelerate the transition

---

[76] A "DX rating" is DoD's "highest national defense urgency priority rating symbol."

[77] Government Accountability Office, *Testimony Before the House Armed Services Committee, Defense Acquisition Reform Panel: Defense Acquisitions: Rapid Acquisition of MRAP Vehicles; Statement of Michael J. Sullivan, Director Acquisition and Sourcing Management*, GAO-10-155T, October 8, 2009, p. 6.

[78] GAO, October 8, 2009.

of high-potential science and technology projects into operationally useful products in the execution years.[79]

Through various small reorganizations and a name change, the RRTO is now under the Director of Rapid Fielding in DDR&E.

RRTO's mission is to

> partner with DoD offices, other government agencies, industry and academia to counter emerging and anticipated threats; and respond to validated joint urgent needs by accelerating the development and fielding of affordable, sustainable traditional and non-traditional capabilities for the Warfighter.[80]

The office focuses on 6- to 18-month projects without focusing on a particular threat. RRTO also tries to seek technology solutions anywhere in DoD and even in other parts of the federal government.

The final two processes discussed here are Army processes that were set up to evaluate the future use of rapidly acquired ad hoc Army capabilities. These processes determine which of the ad hoc capabilities should become Army-wide programs of record. They also evaluate the capabilities to see which should be sustained, which means that the capabilities will be in place and funded for as long as specific parts of the Army need the capabilities. Finally, if there is no longer a need for a specific capability, then the capability will be terminated.

Capabilities Development for Rapid Transition (CDRT) Process Addresses Transition Decision

The Capabilities Development for Rapid Transition (CDRT) process began in October 2004. It was originally called Spiral to the Army at Large. In 2004, the Vice Chief of Staff, Army (VCSA) asked TRADOC to identify all of the nonstandard equipment that was being procured rapidly by the Army to see which of the enduring capabilities would be worth expediting a transition to program of record status. Nonstandard equipment generally can be from the Rapid Equipping Force, the Rapid Fielding Initiative, JIEDDO, Program Manager requests, COTS purchases, and any other effort that is a way of procuring and fielding equipment outside of the normal acquisition process.

---

[79] "OSD RDT&E BUDGET ITEM JUSTIFICATION (R2 Exhibit): RDTE, Defense Wide BA 03: 0603826D8Z - Quick Reactions Special Projects (QRSP)," Defense Technical Information Center (DTIC), February 2008, p. 222.

[80] "Rapid Fielding Directorate: Rapid Reaction Technology Office," *Rapid Fielding* Directorate, July 2010.

The CDRT process is managed by TRADOC's Army Capabilities Integration Center (ARCIC). The CDRT process

> begins with ARCIC's Asymmetric Warfare Division and the Rapid Equipping Force identifying potential systems, vetting that list with the various COCOM and Army Commands, soliciting their views and additional candidates, and running the recommendations through a selection process that includes TRADOC, those same Army commands and COCOM, and the DA staff. At each step of the process, TRADOC and [Headquarters, Department of the Army] provide input. Some selected systems will be fielded to the Army at large . . . while others will have limited application . . . [and] will remain in theater.[81]

Figure A.4 illustrates the general CDRT process.

Figure A.4. Capabilities Development for Rapid Transition Process

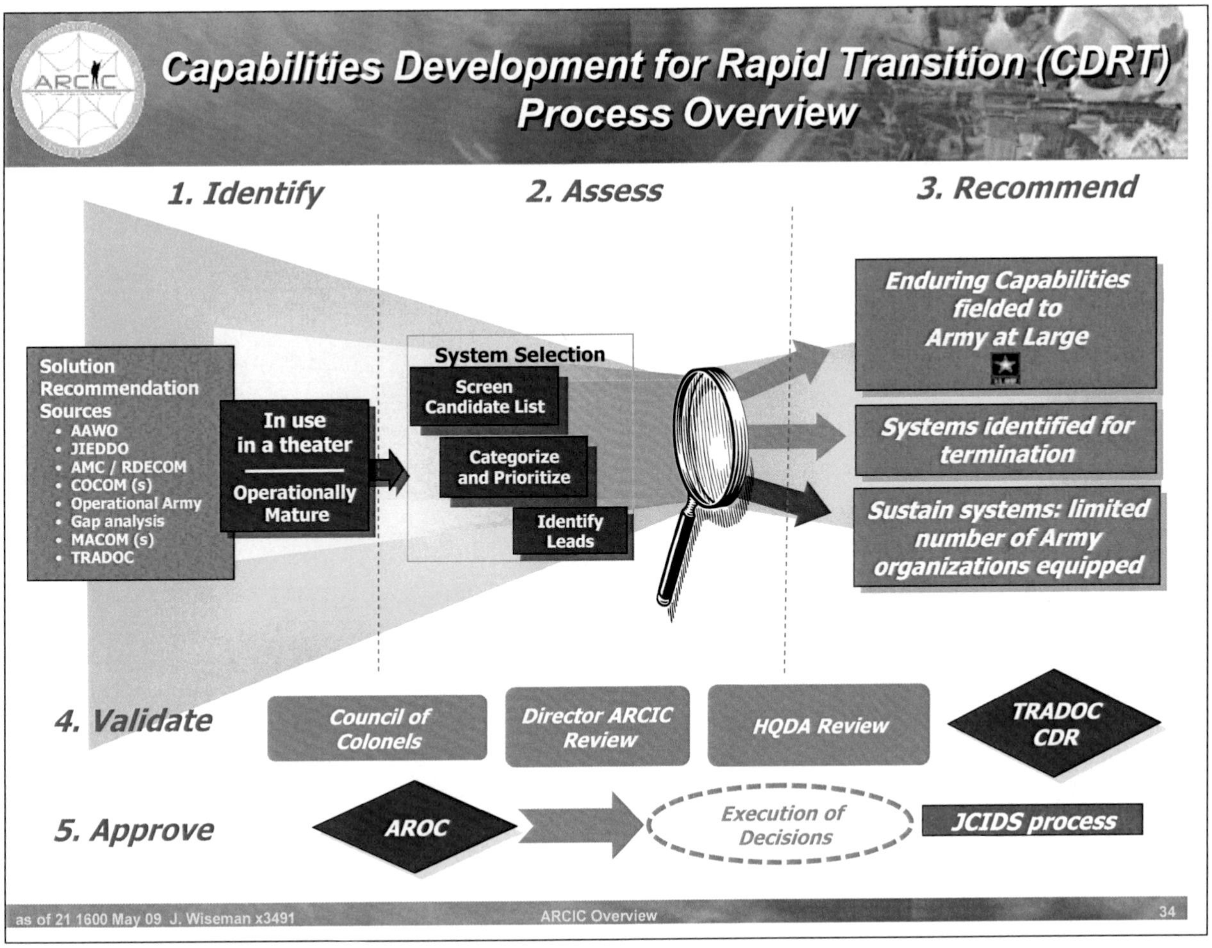

---

[81] "Army Capabilities Integration Center," U.S. Army Training and Doctrine Command, June 13, 2008, p. 10.

If a program is successfully used in theater and approved for the CDRT list, it may bypass capabilities-based assessment steps, which include functional area analysis, functional needs analysis and functional solutions analysis.[82]

Those chosen to become PORs would transition to the Army's normal acquisition system and become available Army-wide. The criteria for an enduring capability include:

- Has wider use in the Army than just current operations in Iraq or Afghanistan;
- Must be successfully used in theater for at least 120 days;
- Not be a current POR ;
- Be operationally mature;
- Has a current operational assessment where the minimal requirement is the completion of the Army Test and Evaluation Command's (ATEC) Capabilities and Limitations report;
- Be capable of being produced with little modification;
- Solve a current capability gap; and
- Be applicable to future requirements in the Army.[83]

Other programs where the capabilities did not meet the above criteria, but would currently fill some operational need in Iraq or Afghanistan, would be labeled as sustain (previously niche). The sustain capabilities are ones that are important for current operations, but may not be necessary in the future. These capabilities generally are funded through supplemental funds and are re-evaluated in future iterations of CDRT based on operational need.

If a capability is no longer needed by the Army, then the CDRT process decides to terminate the capability in the future. The Army cannot always make a binding decision on whether a capability should be terminated or become a POR, so the capability would be listed as sustain and then would be re-evaluated in a later iteration to see if the program should continue or be terminated.

CDRT originally started as an annual process. In 2008 (iteration 5), it became semiannual. In May 2009 (iteration 6), VCSA General Peter Chiarelli made the process quarterly in order to accelerate the evaluation of the various capabilities. As of January 2010 (iteration 8):

---

[82] Kate Brannen, "Capabilities Development for Rapid Transition Evolves: Army Eyes Slate of Promising Efforts for 'Program of Record' Status," *Inside the Army*, October 6, 2008.

[83] Brannen, "Capabilities Development for Rapid Transition Evolves," 2008.

- 470 capabilities (material and nonmaterial) were considered;
- 51 were selected as enduring; and
- 117 were selected for termination.[84]

Army's Immediate Warfighter Needs (IWN) Process

Outside of the CDRT process, the Army uses its Immediate Warfighter Needs (IWN)[85] process to transition capabilities to lifecycle management. The Army Asymmetric Warfare Office (AAWO) initiated the IWN process in 2007, to formalize the transition of projects from JIEDDO to the Army. Shortly thereafter, the process was opened to non-JIEDDO initiatives. It is a more informal process than CDRT and is used infrequently. The process has only been used for approximately 40 to 50 items. If an item does not go through the CDRT process, it can utilize the IWN process instead.

The IWN process assigns a capability to either initial program management or lifecycle support to the Army Acquisition Executive (AAE) (within the ASA(ALT) PEO/PM structure) or to the Army Materiel Command (AMC). The capability can come from a number of sources, including CDRT, REF, TRADOC, RDECOM, and other organizations. Before a transfer can take place, there are two main issues to be worked out: there needs to be a validated ONS/JUONs, and the funding profile must be set up. After these two major issues are worked out, the PEO/PM or AMC needs to agree to take responsibility for the item.

IWN is different from CDRT because the intention of CDRT is to look for PORs, while the IWN process is simply placing capabilities that will never be more than sustain or niche capabilities in lifecycle management/sustainment. These items typically fulfill short-term needs of units within the Army, rather than the entire Army.

Joint/Army Rapid Capabilities and Materiel Developments Initiatives Work Together

The rapid capabilities and materiel developments initiatives mentioned in this appendix work together to fulfill urgent warfighter needs. In this section we illustrate that relationships exist between various organizations or processes based on process flow charts or anecdotal evidence. We do not attempt here to establish how well or to what extent these processes work together. There are relationships between many of these

---

[84] "Capabilities Development for Rapid Transition (CDRT): Previous Iteration Status Tables," Accelerated Capabilities Division, Accelerated and Capabilities Development Directorate, Army Capabilities Integration Center (ARCIC), TRADOC; Current and Future Warfighting Capabilities Headquarters, Department of the Army G-3/5/7 (DAMO-CI), January 13, 2010.

[85] The term "Immediate Warfighter Needs" is used by JRAC to identify needs (discussed earlier) and by the Army for its transition process.

processes at both the Joint and Army levels. Some of the stronger linkages are discussed below.

Operational Needs Statement (ONS), Immediate Warfighter Need (IWN), Joint Urgent Operational Need (JUON), Army Requirements and Resourcing Board (AR2B), and Joint Rapid Acquisition Cell (JRAC)

The processes that identify needs—ONS, Joint IWN, and JUON—work together with the processes that prioritize the needs.[86] After an Army ONS is identified, the Army Requirements and Resourcing Board certifies the need. Also, the Joint Rapid Acquisition Cell (JRAC) reviews JUONs and then designates the more critical IWNs from the larger set. After prioritizing the needs, both the AR2B and the JRAC make sure that the needs are funded and then procured by a process like REF or JIEDDO.

Rapid Equipping Force (REF), Joint Rapid Acquisition Cell (JRAC), Mine Resistant Ambush Protected (vehicle) Task Force (MRAP TF), and Joint Improvised Explosive Devices Defeat Organization (JIEDDO)

A different type of relationship exists between REF, JRAC, MRAP TF, and JIEDDO. REF, JRAC, MRAP TF, and JIEDDO are independent of each other, but share information when necessary.

> The JIEDDO works in partnership with REF . . . in preparing Soldiers and leaders to face the pervasive IED threat in the current operating environment.[87]

JRAC works with all three organizations because each JUON needs a service/JIEDDO sponsor that will provide procurement and sustainment. JRAC would then monitor the progress of these organizations in fulfilling a particular need. All of these organizations also have direct linkages to the urgent needs processes: ONS, IWN, and JUON. A valid urgent needs statement is required for a need to be funded and procured.

Quick Reaction Fund (QRF)/Rapid Reaction Fund (RRF)/Rapid Reaction Technology Office (RRTO)/Joint Rapid Acquisition Cell (JRAC)

QRF and RRF are funds that are managed by the Rapid Reaction Technology Office. In this case, an organizational relationship exists. RRTO decides which fund will be used for emerging needs. The Rapid Reaction Fund (RRF) provides funding to projects that focus on counterterrorism or counterinsurgency. It is a joint fund that is managed by the Rapid Reaction Technology Office (RRTO). An organizational linkage also exists

---

[86] ONS/JUONs/IWNs are priorities by definition; however, we have not seen any actual evidence of prioritization of those needs. This does not mean that evidence does not exist, but rather that it is not publicly available.

[87] Bob Keenan, "Capabilities Development and System Acquisition Management: 2010 Executive Primer," *Army Force Management School (AFMS)*, Version 15, May 2010, p. 90.

between RRTO and JRAC because both offices report to the Director of Rapid Fielding in the office of the Director, Defense Research and Engineering.

Capabilities Development for Rapid Transition (CDRT) and Other Army Processes

The CDRT process maintains relationships to nearly all of the Army rapid acquisition processes discussed in this section because it is the main transition process established by the Army. As was discussed earlier, the CDRT process reviews systems acquired in response to urgent needs to see if any of them should become programs of record, which would allow the capabilities to be fielded throughout the Army and sustained as part of the Army's normal lifecycle management process. Each program that enters the CDRT process must have a valid ONS or JUON. Then, the CDRT process reviews capabilities that were approved by the AR2B and that were procured by REF, RFI, or JIEDDO.

CDRT works directly with the REF and RFI processes. This is evident in the output from the CDRT process. Cross-checking the REF and RFI programs with those that have gone through the CDRT process provides a list of more than 27 programs that went through the CDRT process.

In addition, CDRT works together with the Army IWN process; IWN offers, in its current form, a second look at non-PORs exiting the CDRT process that were not selected to transition to POR status. Both are Army transition processes that require an official ONS in order to begin the transition process. The IWN receives some of its candidates from the CDRT process if the capabilities are not a good fit for that process. It also receives some candidates from REF and JIEDDO.

Challenges Facing Rapid Capabilities and Materiel Developments Initiatives

Most of the above processes have been in use for over five years. The processes have been effective in providing the warfighter with rapidly acquired capabilities; however, the various processes are now coming under closer scrutiny by DoD, Congress, the services, and the GAO as DoD tries to develop an overall path for the future of rapid acquisition. Some common challenges are now visible with respect to these processes. JIEDDO and MRAP, in particular, have received a lot of scrutiny from Congress, the GAO, and others because these efforts have received a significant amount of funding to counter IEDs in both Iraq and Afghanistan.

Some of the most recent literature on MRAP found that the effort has experienced a variety of challenges along with its major acquisition successes. The GAO found that the MRAP program was successful because DoD made the program its highest priority; provided a significant amount of funding; and provided quick access to critical materials. However, after the MRAPs were fielded, the program experienced design-related challenges. The designs of the vehicles had to be modified to deal with reliability, mobility, and safety issues. Other challenges to the MRAP program included budgeting and sustainment issues. It is unclear exactly how MRAPs will fit into DoD's longer-term operational plan, which makes it difficult to establish a budget and to estimate sustainment costs over time.[88] Finally, the Cost Analysis Initiative Group (CAIG)[89] called attention to some cost estimating issues related to the MRAP program. Due to the urgency of the need, cost estimating was waived for the MRAP program. Accurate cost estimates have been found to be critical to the long-term success of a program. The CAIG explained that it needed to "estimate production costs without the benefit of prototype costs and testing experience."[90] It was also concerned about the significant roles that contractors played in the program, which was uncommon in other traditional acquisition programs. Finally, the CAIG was concerned about data collection given that this was an urgent effort.[91] These challenges are a direct consequence of the rapid acquisition process used to acquire MRAPs.

The JIEDDO effort, like the MRAP effort, has received a lot of scrutiny. Congress mandated that the GAO conduct several studies on JIEDDO's management and operations. One GAO study from October 2009 found that JIEDDO is experiencing several management and operations challenges. First, JIEDDO does not have full visibility over all counter-IED efforts in DoD, because there is no database that tracks all efforts throughout DoD. This creates the challenge of trying to control duplication of DoD-wide IED efforts. NEXT, JIEDDO is having trouble with transitioning IED initiatives to the services. Consequently, the services often rely on Overseas Contingency Operation (OCO) funding. This causes an uncertain funding future for counter-IED initiatives outside of JIEDDO.[92]

In a more recent GAO report, the GAO found that JIEDDO "has not yet developed a means for reliably measuring the overall effectiveness of its efforts and investments to combat IEDs."[93] The GAO points out that there are some factors given the unique operational environment that are limiting the organization's ability to collect data. Also, JIEDDO has found it difficult to fully adhere to its review and approval process for

---

[88] GAO, October 8, 2009, summary section.

[89] CAIG is now part of the recently created Cost Analysis and Program Evaluation (CAPE) office, which replaced the Program Analysis and Evaluation (PA&E) office.

[90] Walt Cooper, "Rapid Acquisition Strategies: Implications for Cost Analysts—The MRAP Example," Office of the Secretary of Defense (Cost Analysis Improvement Group), February 19, 2009, p. 5.

[91] Cooper, February 19, 2009, p. 5.

[92] Government Accountability Office, *Warfighter Support: Actions Needed to Improve Visibility and Coordination of DOD's Counter-Improvised Explosive Device Efforts*, GAO-10-95, October 2009, summary section.

[93] Government Accountability Office, *Warfighter Support: Actions Needed to Improve the Joint Improvised Explosive Device Defeat Organization's System of Internal Control*, GAO-10-660, July 2010, summary section.

counter-IED initiatives that were being developed. This challenge has resulted in a lack of transparency for how funds are being spent. Finally, JIEDDO was found to have some significant internal control system weaknesses that continue to be problematic for the organization. The GAO believes that these problems exist because of

> a lack of sustained management attention in following through with corrective actions; challenges with retention and expertise of personnel; and a lack of sufficient acquisition expertise with breadth and depth to understand the programs.[94]

JRAC, another joint organization, has talked about the challenges the organization faces. Specifically, the organization identified finding funding for IWNs as problematic. In addition, JRAC found that it was having difficulty keeping track of "sponsor actions, schedules, and commitments once funding has been provided."[95]

The Army's AR2B is an organization with a similar role to the JRAC. One of the challenges that the AR2B faces is similar to the JRAC and most other urgent-needs processes and organizations. The AR2B has found that funding ONS is difficult. AR2B funds most of its efforts through supplemental funding and budget tradeoffs because it does not have a base budget. This could be problematic if supplemental funding is reduced significantly.[96]

The GAO released a report in April 2010 that called out challenges that have generally been seen in urgent needs processes. The GAO found that DoD cannot judge how well its urgent needs processes are working "because it has not established an effective management framework for those processes."[97] The GAO came to this conclusion after finding the following evidence:

> DOD's guidance for its urgent needs processes is dispersed and outdated . . . DOD guidance does not clearly define roles and responsibilities for implementing, monitoring, and evaluating all phases of those processes or incorporate all of the expedited acquisition authorities available to acquire joint urgent need solutions . . . Data systems for the processes lack comprehensive, reliable data for tracking overall results and do not have standards for collecting and managing data . . . In addition, the joint process does not include a formal method for feedback to inform joint leadership on the performance of solutions.[98]

---

[94] GAO, July 2010, summary section.

[95] William E. Beasley (Director (Acting)), "Overview of Joint Rapid Acquisition Cell for the Department of Defense Cost Analysis Symposium," Office of the Secretary of Defense, Joint Rapid Acquisition Cell, February 19, 2009, p. 17.

[96] Dabolt,. pp. 5–6.

[97] GAO, April 2010, summary section.

[98] GAO, April 2010, summary section.

In addition to the above issues, the GAO found that some personnel involved in examining urgent needs do not have proper training in dealing with urgent needs processes, which can lead to difficulties and delays in processing requests in theater. Another challenge that GAO found involved funding. There are widespread problems among urgent needs processes in DoD with regard to obtaining funding for urgent needs. Difficulty in obtaining funding was

> the primary factor that increased the amount of time needed to field solutions. Funding has not always been available for joint urgent needs in part because the Office of the Secretary of Defense (OSD) has not assigned primary responsibility for implementing the department's rapid acquisition authority.[99]

Finally, the use of immature or complex technologies has resulted in delays in fulfilling urgent needs among various urgent needs processes throughout DoD.

The Defense Science Board, like the GAO, was tasked with evaluating urgent needs processes throughout DoD. The result was a report that was released in July 2009. The report addresses various issues regarding urgent needs processes. Of particular interest to this study are the challenges that it called out throughout its report. First, the DSB noted that one of the initial challenges is taking a need that comes in from the field and categorizing the need as either something that should be acquired through the acquisition system or something that should be filled by a logistics function.

> Evaluating a true capability gap in an effective and systematic way . . . involve[s] operations research and system analysis (ORSA) and analysis of alternatives . . . This ORSA step . . . is missing in many current rapid acquisition processes.[100]

The Army has had some difficulty with this task. Since 2003, 90 percent of the ONS have been for "distribution and redistribution of inventory items" rather than for finding a solution for a need through equipment that is not already at the Army's disposal.[101]

This DSB study also found that many of the ad hoc task forces and programs that currently exist in the urgent needs framework in DoD may not have "impetus or adequate advocacy beyond the war"[102] because there still has not been a lot of serious institutional commitment. These processes will likely experience budget battles with programs of record in the future unless funding is done differently.[103] There are some additional issues that arise with there being ad hoc processes that are fulfilling urgent needs. It is problematic that these processes rely on "learning on-the-job with little emphasis on

---

[99] GAO, April 2010, summary section.

[100] DSB, July 2009), July 2009, p. viii.

[101] DSB, July 2009, p. 12.

[102] DSB, July 2009, p. ix.

[103] DSB, July 2009, p. ix.

support, training, and sustainment."[104] One of the main ways that urgent needs are being fulfilled is through COTS. The DSB study noted that "DoD acquisition personnel have limited visibility of emerging technologies."[105] This is mainly due to the fact that DoD acquisition stresses that center on major platforms rather than capabilities.

Other shortfalls in the urgent needs process that were raised by the DSB include:

> There is no consistent system in place that documents total time and cost to satisfy each need statement. In general, the task force observed uneven tracking of field performance of the capability implemented or materiel delivered, and only ad hoc assessment of how original need was addressed. Further, there was little coordination among the Services to determine if needs were shared and solutions could be more widely applied. Most importantly, methods to assess sustainment needs or future-year costs have lagged implementation, with alarming consequences for future DoD budgets.[106]

Each rapid acquisition process within DoD has some challenges that need to be overcome, but generally, the warfighters have been receiving what they need. Given that the processes are mostly under ten years old, and that there has been no systematic process to evaluate their effectiveness, we recommend the Army support the development of metrics to characterize the efficiencies conveyed by the processes, and their effectiveness. Especially if DoD and the Army finish arrangements to institutionalize rapid acquisition in the longer term, this data and the associated metrics will provide policy makers with a better understanding of which aspects of which processes should be institutionalized, and which should be merged or eliminated to support conduct of rapid acquisition in the future.

---

[104] DSB, July 2009, p. 6.

[105] DSB, July 2009, p. 4.

[106] DSB, July 2009, p. 19.

(This page is intentionally blank)

Bibliography

Authors' note: Some of the journals cited in this bibliography can be accessed through InsideDefense.com, a subscription-only website. Searchable journals included on this site include *Inside the Pentagon, Inside the Army, Inside the Navy, Inside the Air Force, Homeland Defense Watch, Inside Missile Defense, Defense Alert, Defense Plus,* and *Defense Information & Electronics Report (DIER).*

"A New Approach for Delivering Information Capabilities in the Department of Defense," Office of the Secretary of Defense, draft, July 22, 2010.

"Army Capabilities Integration Center," U.S. Army Training and Doctrine Command, June 13, 2008, p. 10. As of June 14, 2010: http://www.arcic.army.mil/Briefings/ARCIC%20Overview_0613Jun08.ppt#291,10

Army Strong: Equipped, Trained and Ready: Final Report of the 2010 Army Acquisition Review, January 2011. As of July 2013: http://www.rdecom.army.mil/EDCG%20Telecoms/ Final%20Report_Army%20Acq%20Review.pdf

Barbaris, Roxanne, and Christine Callanan, *United States Army Contingency Contracting Operations: Emerging Roles, Procedures, and Challenges Facing Contracting Professionals,* Joint Applied Project, Naval Postgraduate School, September 2008.

Beasley, William E., "Overview of Joint Rapid Acquisition Cell," briefing done for the Department of Defense Cost Analysis Symposium, February 19, 2009.

Benedict, Lieutenant Colonel Leith A., *Converging the Networks,* USAWC Strategy Research Project, Carlisle, PA: U.S. Army War College, March 26, 2007.

Bertuca, Tony, "Army Announces Acquisition Review to Speed Up, Streamline Process," *Inside the Pentagon,* June 3, 2010.

Blakeman, MAJ Seth T., MAJ Anthony R. Gibbs, MAJ Jeyanthan Jeyasingam, *Study of the Mine Resistant Ambush Protected (MRAP) Vehicle Program as a Model for Rapid Defense Acquisitions,* Acquisition Research Sponsored Report Series, NPS-AM-08-132, Monterey, CA: Naval Postgraduate School, December 3, 2008.

Boessenkool, Antonie, "DoD IT Procurement Too Slow: Cartwright," *DefenseNews.com,* March 4, 2009.

Bonomo, James, Donald Prosnitz, Douglas Shontz, and Shara Williams, *Fostering Innovation in Information Technology at the FBI,* Santa Monica, CA: RAND Corporation, PM(L)-2421-ISE, January 2008. Not available to the general public.

Brannen, Kate, "Army Eyes Slate of Promising Efforts for 'Program of Record' Status," *Inside the Army*, Vol. 20, No. 40, October 6, 2008. Available at InsideDefense.com, a subscription-only website.

Brannen, Kate, "Chiarelli: Rapid Equipping Force Provides Great Model for Reform," *Inside the Army*, Vol. 21, No. 4, February 2, 2009. Available at InsideDefense.com, a subscription-only website.

Brannen, Kate, "Dempsey: Rapid Pace of Change Means Shorter Development Time Lines," *Inside the Army,* Vol. 21, No. 51, December 28, 2009. Available at InsideDefense.com, a subscription-only website.

Brannen, Kate, "Pentagon Set to Launch Independent Study of Meads Program," *Inside the Army*, Vol. 21, No. 21, June 1, 2009. Available at InsideDefense.com, a subscription-only website.

Brannen, Kate, "Rapidly Developed 'Green Dart' and 'TIGR' Become Programs of Record," *Inside the Army*, Vol. 21, No. 2, January 19, 2009. Available at InsideDefense.com, a subscription-only website.

Bryant, Colonel Thomas H., "Rapid Fielding Initiative Overview to the 2005 Acquisition Senior Leaders and AMC Commanders Conference," briefing, August 23, 2005. As of June 30, 2010:
http://asc.army.mil/docs/briefings/slc_2005/RFI_Overview_for_ACQ_Seminar_Working_Group.pdf

Buhrkuhl, Robert L., "When the Warfighter Needs It Now," *Defense AT&L,* November–December 2006.

Business Executives for National Security, "Getting to Best: Reforming the Defense Acquisition Enterprise," Defense Acquisition Archives, July 2009. As of July 2013:
http://www.bens.org/document.doc?id=44

"Buying Commercial: Gaining the Cost/Schedule Benefits for Defense Systems," Report, Defense Science Board Task Force on Integrating Commercial Systems into the DOD, Effectively and Efficiently, February 2009.

"Capabilities Development for Rapid Transition (CDRT): Previous Iteration Status Tables," Accelerated Capabilities Division, Accelerated and Capabilities Development Directorate, Army Capabilities Integration Center (ARCIC), TRADOC; Current and Future Warfighting Capabilities Headquarters, Department of the Army, G-3/5/7 (DAMO-CI), January 13, 2010.

"Capabilities Development for Rapid Transition (CDRT) Information Briefing, Previous Iteration Status Tables," Headquarters, Department of the Army G-3/5/7 (DAMO-CI), October 2, 2009.

Castelli, Christopher J., "DOD Testing Chief Launches Efforts to Quickly Field, Improve Systems," January 28, 2010.

Censer, Marjorie, "Army Science Board Recommends DOD Establish Innovation Cell," *Inside the Army,* February 1, 2010. Available at InsideDefense.com, a subscription-only website.

Censer, Marjorie, "Ford: Army Must Eventually Merge Rapid, Traditional Acquisition," *Inside the Army,* September 15, 2008. Available at InsideDefense.com, a subscription-only website.

Censer, Marjorie, "JIEDDO Acquisition Director Seeks Improved Transparency, Clear Guidance," *Inside the Army,* April 12, 2010. Available at InsideDefense.com, a subscription-only website.

Censer, Marjorie, "JIEDDO Holds Monthly Meetings With Service Officials, Opens Database," *Inside the Army,* April 5, 2010. Available at InsideDefense.com, a subscription-only website. Available at InsideDefense.com, a subscription-only website.

Censer, Marjorie, "JIEDDO to Provide Early Notification Before Transitioning Initiatives," *Inside the Army,* May 28, 2009. Available at InsideDefense.com, a subscription-only website.

Censer, Marjorie, "Metz: JIEDDO Faces 'Altogether Different Challenge' in Afghanistan," *Inside the Army,* November 2, 2009. Available at InsideDefense.com, a subscription-only website.

Censor, Marjorie, and Jason Sherman, "Pentagon IG: Defense Department 'Likely Overpaid' for MRAP Vehicles," *Inside the Army,* February 2, 2009. Available at InsideDefense.com, a subscription-only website.

Clark, Colin, "iPhone Likely Loser for DoD Biz," *DoD Buzz,* June 14, 2010.

"Command, Control, Communications, Computers, and Intelligence (C4I) Systems," *Army,* October 2007.

Concepcion, Diane, "WIN-T Overview and Cross Incremental Supportability," Warfighter Information Network–Tactical Briefing, Team C4ISR, Fort Monmouth, May 13, 2008.

Cooper, Walt, "Rapid Acquisition Strategies: Implications for Cost Analysts—The MRAP Example," Office of the Secretary of Defense (Cost Analysis Improvement Group), February 19, 2009. Available at InsideDefense.com, a subscription-only website.

Crain, William Forrest, "Army Equipping Strategy Study, Terms of Reference," briefing, Army Materiel Command, July 28, 2009.

Dabolt IV, CPT John H., "Army Requirements and Resourcing Board: Rapid Reaction in an Era of Persistent Conflict," *The Oracle*, Vol. 4, 4th Quarter FY 2008. As of June 9, 2010:
http://www.fa50.army.mil/pdfs/OracleFA50News_Vol4_4Q_FY08.pdf

Davidson, Joshua, "PEO C3T—Bringing the Future to the Present Fight," *Army AL&T,* October–December 2008.

Defense Acquisition University (DAU), Acquisition Community Connection website, "Rapid Acquisition Authority to Respond to Combat Emergencies," January 14, 2005. As of January 1, 2013:
https://acc.dau.mil/CommunityBrowser.aspx?id=18589

Defense Science Board (DSB), *Fulfillment of Urgent Operational Needs*, Washington, D.C.: Office of the Under Secretary of Defense for Acquisition, Technology, and Logistics, July 2009. As of June 9, 2010:
http://www.ndia.org/Advocacy/Resources/Documents/LegislativeAlerts/DSB_Urgent_Needs_Report_7_15_2009.pdf

Department of Army, Memorandum, "Procedures for Transfer of Rapidly Equipped Initiatives/Products/Systems to Program Executive Offices (PEOs) for Life-Cycle Management," October 24, 2008, signed by Dean G Popps, Principal Deputy Assistant Secretary of the Army (Acquisition, Logistics and Technology).

Department of Defense Policies and Procedures for the Acquisition of Information Technology, Report of the Defense Science Board Task Force, Office of the Under Secretary of Defense for Acquisition, Technology, and Logistics, March 2009. As of July 2013:
http://www.acq.osd.mil/dsb/reports/ADA498375.pdf

Eidson, Tech. Sgt. Shad, *"Battlespace Command, Control Center Protects Region," Air Force Print News,* March 13, 2009.

Erwin, Sandra I., "Army's Vice Chief: 'We Have to Speed Up How We Buy Things,'" *National Defense,* October 2009.

Fiedler, LTC David M., "MBITR Communications—Power in Your Pocket," *Army Communicator,* Summer 2005.

Fogg, Glenn, "How to Better Support the Need for Quick Reaction Capabilities in an Irregular Warfare Environment: Quick Reaction and Rapid Reaction Funds," Rapid Reaction Technology Office (RRTO), April 21, 2009. As of June 30, 2010:
http://www.dtic.mil/ndia/2009science/Fogg.pdf

Fuller, COL(P) Peter N., "Rapid Acquisition—Developing Processes That Deliver Soldier Materiel Solutions Now," United States Army Acquisition Support Center, February 2008. As of June 10, 2010:
http://www.usaasc.info/alt_online/article.cfm?iID=0802&aid=15

Gansler, Jacques S., and William Lucyshyn, "Commercial-Off-The-Shelf (COTS): Doing It Right," Center for Public Policy and Private Enterprise, School of Public Policy, University of Maryland, September 2008.

Glass, Jon W. "Rush Order: $3 Billion Worth of Aircraft, Sensors Leaves Need for Reform, Analysts Warn," *C4ISRJournal.com*, March 1, 2009. As of March 6, 2009:
http://www.c4isrjournal.com/story.php?F-3910017

Gonzales, Daniel, Eric Landree, John Hollywood, Steven Berner, and Carolyn Wong, *Navy/OSD Collaborative Review of Acquisition Policy for DoD C3I and Weapon Programs*, Santa Monica, CA: RAND Corporation, DB-528-NAVY/OSD, 2007.
http://www.rand.org/pubs/documented_briefings/DB528.html

Gould, Joe, "DOD Acquisition Chief Approves WIN-T Program Baseline, with Caveats," *Inside the Army,* June 1, 2009.

Gould, Joe, "New CIO/G-6 Official Calls for More Bandwidth in Combat Theaters," *Inside the Army,* October 26, 2009.

Government Accountability Office, *An Analysis of the Special Operations Command's Management of Weapon System Programs*, GAO-07-620, June 2007.

Government Accountability Office, *Department of Defense Needs Framework for Balancing Investments in Tactical Radios*, GAO-08-877, August 2008.

Government Accountability Office, *Defense Technology Development: Management Process Can Be Strengthened for New Technology Transition Programs*, GAO-05-480, June 17, 2005.

Government Accountability Office, *DOD's Requirements Determination Process Has Not Been Effective in Prioritizing Joint Capabilities*, GAO-08-1060, September 2008.

Government Accountability Office, *Testimony Before the House Armed Services Committee, Defense Acquisition Reform Panel: Defense Acquisitions: Rapid Acquisition of MRAP Vehicles; Statement of Michael J. Sullivan, Director Acquisition and Sourcing Management*, GAO-10-155T, October 8, 2009. As of June 30, 2010:
http://www.gao.gov/new.items/d10155t.pdf

Government Accountability Office, *Warfighter Support: Actions Needed to Improve Visibility and Coordination of DOD's Counter-Improvised Explosive Device Efforts*, GAO-10-95, October 2009. As of June 30, 2010:
http://www.gao.gov/new.items/d1095.pdf

Government Accountability Office, *Warfighter Support: Actions Needed to Improve the Joint Improvised Explosive Device Defeat Organization's System of Internal Control*, GAO-10-660, July 2010. As of June 30, 2010:
http://www.gao.gov/new.items/d10660.pdf

Government Accountability Office, *Warfighter Support: DOD's Urgent Needs Processes Need a More Comprehensive Approach and Evaluation for Potential Consolidation*, March 2011, GAO-11-273. As of July 2013:
http://www.gao.gov/new.items/d11273.pdf

Government Accountability Office, *Warfighter Support: Improvements to DOD's Urgent Needs Processes Would Enhance Oversight and Expedite Efforts to Meet Critical Warfighter Needs*, GAO-10-460, April 2010. As of June 9, 2010:
http://www.gao.gov/new.items/d10460.pdf

Greene, Colonel Harry, Larry Stotts, Ryan Paterson, and Janet Greenberg, "CPOF: A Successful Transition of an S&T Initiative to a POR," Defense Advanced Research Projects Agency, *Defense AR Journal*, Vol. 53, January 2010.

Headquarters, Department of the Army (HQDA), Deputy Chief of Staff G-3/5/7, REF Annex to the 2013 Army Posture Statement, December 2012.

"Individual Equipment and Weapons," Association of the United States Army, *Army Magazine*, October 2009. As of June 11, 2010:
http://www.ausa.org/publications/armymagazine/archive/2009/10/Documents/weapons_individual.pdf

"Joint Network Node (JNN)," The Computerworld Honors Program Case Study, U.S. Army, 2006.

"Joint Network Node Program: Datapath Delivers Reliable, High-Bandwidth Communications to the Front Line," Datapath Case Study, 2007.

Johnson, CPT Byron G., "JNN: Reorganizing to Bridge Gaps in Communications," *Army Communicator,* Summer 2005.

Jones, Walter, "Case Analysis of the U.S. Army Warfighting Rapid Acquisition Program: Bradley Stinger Fighting Vehicle—Enhanced Weapon System," thesis, Monterey, CA: Naval Postgraduate School, June 1996.

Jozwiak, Edward, "Rapid Equipping Force (REF) Info: Q. Request for information on REF (UNCLASSIFIED)," email to Shara Williams, February 25, 2013.

Keenan, Bob, "Capabilities Development and System Acquisition Management: 2010 Executive Primer," Army Force Management School (AFMS), Version 15, May 2010. As of June 9, 2010:
http://www.afms1.belvoir.army.mil/pages/primers/primers.html

Keeton, Ann, "Harris Gets Defense Boost," *Wall Street Journal,* February 17, 2010.

Lovett, Col. Robert A., "Rapid Equipping Force: Streamlined Acquisition Process," U.S. Army Rapid Equipping Force, October 27, 2005. As of June 9, 2010: http://www.dtic.mil/ndia/2005expwarfare/lovett.pdf

Magnus, R., "Marine Corps Order 3900.15B, Marine Corps Expeditionary Force Development System (EFDS)," Commandant of the Marine Corps, Department of the Navy, March 10, 2008.

Metz, LTG Thomas (Director, Joint IED Defeat Organization), "Media Roundtable," Joint Improvised Explosive Device Defeat Organization, April 30, 2008.

Muñoz, Carlo, "Joint Staff Kicks Off Major Review of JCIDS, JUONs Requirements Process," *Inside the Air Force,* August 20, 2010.

Muñoz, Carlo, "Official: Criticisms of JIEDDO, ISR Task Force Could Lead to Oversight Overkill," *Inside the Air Force,* May 21, 2010.

Muñoz, Carlo, "Riley: Consolidation of Rapid Acquisition Processes Is Unnecessary," *Inside the Pentagon,* October 30, 2008.

Muñoz, Carlo, "Senate Warns DOD on Flexibility Granted to JIEDDO, ISR Task Force," *Inside the Navy,* May 17, 2010.

Muñoz, Carlo, "Gates: Pentagon Shutting Down MRAP, ISR Task Forces," *Inside the Pentagon,* June 17, 2010.

National Research Council, *Achieving Effective Acquisition of Information Technology in the Department of Defense*, Washington, D.C.: The National Academies Press, 2010. As of August 2013: http://www.nap.edu/openbook.php?record_id=12823

"Nonstandard Equipment and Materiel Maintenance and Sustainment," Department of the Army, Pamphlet 750-X.

Office of the Deputy Chief Management Officer Representative, "Review of Acquisition Processes for Rapid Fielding of Capabilities in Response to Urgent Operational Needs" (draft version), Department of Defense, October 21, 2012.

Olson, Admiral Eric T., "Statement before the Senate Armed Services Committee on the Posture of Special Operations Forces," March 4, 2008. Not available to the general public.

"OSD RDT&E Budget Item Justification, (R2 Exhibit): RDTE, Defense Wide BA 03: 0603826D8Z—Quick Reactions Special Projects (QRSP)," Defense Technical Information Center (DTIC), February 2008. As of June 30, 2010: http://www.dtic.mil/descriptivesum/Y2009/OSD/0603826D8Z.pdf

Petraeus, General David H., "Adaptive, Responsive, and Speedy Acquisitions," *Defense AT&L,* January–February 2010.

Pincus, Walter, "Defense Dept.'s Special Projects Program Features More Sophisticated Weapons," *Washington Post,* November 10, 2009.

Pollachek, Brandon, "PEO IEW&S—Providing Capabilities to Enhance Warfighter Survivability," *Army AL&T*, October–December 2008.

Popps, Dean G., "Non-Standard Equipment Interim Policy Memorandum," Department of the Army.

Porche, Isaac R., III, Shawn McKay, Megan McKernan, Robert W. Button, Bob Murphy, Kate Giglio, and Elliot Axelband, *Rapid Acquisition and Fielding for Information Assurance and Cyber Security in the Navy*, Santa Monica, Calif.: RAND Corporation, TR-1294-NAVY, 2012. As of October 3, 2013: http://www.rand.org/pubs/technical_reports/TR1294.html

"Rapid Response Process," Air Force Instruction 63-114, Secretary of the Air Force, June 12, 2008.

Rapid Equipping Force, "REF Integrated Priority List," no date. As of February 23, 2013: http://www.ref.army.mil/portal/docs/refprioritylist-RIPL.pdf

"Rapid Fielding Directorate: About RFD," Deputy Assistant Secretary of Defense, Rapid Fielding website, June 2010. As of June 10, 2010: http://www.acq.osd.mil/rfd/about.html#leadership

"Rapid Fielding Directorate: Joint Rapid Acquisition Cell," Joint Rapid Acquisition Cell website, June 2010. As of June 16, 2010: http://www.acq.osd.mil/rfd/jrac.html

"Rapid Fielding Directorate: Rapid Reaction Technology Office," Rapid Fielding Directorate, July 2010. As of July 6, 2010: http://www.acq.osd.mil/rfd/rrto.html

Rasch, Robert A., "Lessons Learned from Rapid Acquisition: Better, Faster, Cheaper?" Carlisle Barracks, PA: Strategy Research Project, U.S. Army War College, December 1, 2011.

Ratnam, Gopal, "Lockheed Gets $399.9 Million Pentagon Order for Blimp," *Bloomberg News*, Bloomberg.com, April 27, 2009. As of April 28, 2009: http://www.bloomberg.com/apps/news?pid=newsarchive&sid=aoMByt6gSFj8

Reiken, Dan, and Chris Gunderson, "PEO C4I Position Paper Re Accelerated IT On-Boarding Via Enhanced Cyber Security Posture," draft, November 4, 2010.

"Report of the Defense Science Board Task Force on the Fulfillment of Urgent Operational Needs," Washington, D.C.: Office of the Under Secretary of Defense for Acquisition, Technology, and Logistics, July 2009.

Riley, Benjamin, "Breaking the Terrorist/Insurgency Cycle," Rapid Reaction Technology Office, Briefing, Overview and Objectives, Washington, D.C.: Department of Defense, 2008.

Robinson, John, "Air & Missile Defense (AMD) Battalion Fire Coordination Cell (FCC) Update," U.S. Army Space and Missile Defense Command Briefing.

Robinson, John, "Air & Missile Defense (AMD) Battalion Fire Coordination Cell (FCC) Update," U.S. Army Space and Missile Defense Command, September 27, 2007.

"Ronald W. Reagan National Defense Authorization Act for Fiscal year 2005, Subtitle B–Amendments to General Contracting Authorities, Procedures, and Limitations, Section 811," 108th Congress of the United States of America, H.R. 4200.

Schwartz, Lieutenant General Norton A., "Rapid Validation and Resourcing of Joint Urgent Operational Needs (JUONs) in the Year of Execution," Chairman of the Joint Chiefs of Staff Instruction, July 15, 2005. As of June 10, 2010:
http://www.dtic.mil/cjcs_directives/cdata/unlimit/3470_01.pdf

Shanker, Tom, "Lightweight Armor Is Slow to Reach the War Zone," *New York Times,* April 17, 2009. As of April 18, 2009:
http://www.nytimes.com/2009/04/18/world/18military.html?hp

Sheikh, Fawzia, "DEE: Pentagon Plans to Alter System of Delivering Warfighter Needs," *Inside the Pentagon,* April 30, 2009.

Sheikh, Fawzia, "DOD Dismisses Call for New Acquisition Agency to Meet Urgent Needs," *Inside the Pentagon,* April 15, 2010.

Sheikh, Fawzia, "New DOD Tool Aims to Quicken Acquisition in Afghanistan, Save Lives," *Inside the Air Force,* November 20, 2009.

Sheikh, Fawzia, "Pentagon Wrestles with Oversight, Speed in IT System Reform," *Inside the Navy,* April 12, 2010.

Sheikh, Fawzia, "Senate Panel Advocates Separate Acquisition System for Urgent Needs," *Inside the Army,* June 14, 2010.

Sheikh, Fawzia, "Wynne-Led Group to Pitch IT System Recommendations to DoD," *Inside the Pentagon,* September 17, 2009.

Sherman, Jason, "DOD Preparing New Rapid Acquisition Policy to Improve War Support," *Inside the Pentagon,* August 5, 2010.

Sherman, Jason, "Gates Renews Call for an Acquisition System More Responsive to Immediate Combat Needs," *Pentagon,* December 5, 2008.

Sherman, Jason, "JFCOM Seeks Support from Obama Team for Enhanced Role in Budget Deliberations," *Defense Alert–Daily News,* June 1, 2009.

Sherman, Jason, "Reorganized PEO Soldier to Improve Focus on Body Armor, Equipment," *Inside the Army,* November 30, 2009.

Sherman, Jason, "Senate Warns DOD on Flexibility Granted to JIEDDO, ISR Task Force," *Inside the Navy,* May 17, 2010.

Sherman, Jason, "Young Directs New Review of Rapid-Acquisition Efforts," *Inside the Pentagon,* January 15, 2009.

"Spiral Technology and Capabilities Development for Rapid Transition to the Army," 2008 Army Posture Statement. As of October 22, 2010: http://www.army.mil/aps/08/information_papers/transform/ Spiral_Technology_and_Capabilities.html

Sprenger, Sebastian, "DSB 'Urgent Needs' Review Kicks Off with One Proposal Already on the Table," *Defense Alert–Daily News,* February 12, 2009.

Sprenger, Sebastian, "Questions Over JIEDDO's Future Touch on Thorny, Familiar Issues," *Inside the Pentagon,* February 12, 2009.

Standifer, Cid, "Legislators Consider MRAP as Model for Rapid Acquisition Programs," *Inside the Army,* October 12, 2009.

Stearns, Scott, "The Rapid Equipping Force: Supporting the American Warfighter," *Infantry Bugler*, National Infantry Association, Fall 2008. As of January 1, 2013: http://www.infantryassn.com/Bugler%20issues/FallBugler2008.pdf

Steele, Jeanette, "Navy Invention Gives Marines Secure Data Connection in Field," *San Diego Union-Tribune,* August 4, 2010.

Stevenson, LTG Mitchell H., "Non-Standard Equipment (NS-E)," briefing, U.S. Army Logistics, Department of the Army, March 18, 2009.

Strayer, Kenneth, Thomas Hoivik, and Susan Page Hocevar, "The Use of Advanced Warfighting Experiments to Support Acquisition Decisions," *Acquisition Review Quarterly,* Fall 2000.

Sullivan, Michael J., "Rapid Acquisition of Mine Resistant Ambush Protected Vehicles," memorandum, U.S. Government Accountability Office, Washington, D.C., July 15, 2008.

Terence, MAJ, "Capabilities Development for Rapid Transition (CDRT) Information Briefing," Headquarters, Department of the Army G-3/5/7 (DAMO-CI), October 2, 2009.

"The U.S. Army's Information Revolution: Delivering Information Dominance to the Warfighter," *Torchbearer National Security Report,* Institute of Land Warfare, Association of the United States Army, August 2006.

"Urgent Reform Required: Army Expeditionary Contracting," Report of the Commission on Army Acquisition and Program Management in Expeditionary Operations, October 31, 2007. As of November 24, 2010: http://www.army.mil/docs/Gansler_Commission_Report_Final_071031.pdf

"Warfighter Information Network–Tactical (WIN-T), Accelerating the Transformation," General Dynamics C4 Systems, 2009.

"Warfighting Capabilities Determination," Army Regulation 71-9, Department of the Army, December 28, 2009. As of October 22, 2010: http://www.fas.org/irp/doddir/army/ar71-9.pdf

Wasserbly, Daniel, "Army Secretary Sees Need for New Rapid Wartime Acquisition System," *Inside the Army,* November 24, 2008.

Wasserbly, Daniel, "Harris Unveils Add-On Radio Mission Modules," *Jane's International Defense Review,* May 27, 2010.

Williams, Shara, Jeffrey A. Drezner, Megan McKernan, Douglas Shontz, and Jerry M. Sollinger, *Rapid Acquisition of Army C2 Systems: Case Studies*, Santa Monica, CA: RAND Corporation, RR-210-A, 2013. Not available to the general public.

"WIN-T—The Building Blocks of Future Communications," *Satellite Evolution,* April/May 2008, pp. 22–26.

Wiseman, J., "Army Capabilities Integration Center," U.S. Army Training and Doctrine Command, May 21, 2009. As of June 14, 2010: http://www.arcic.army.mil/Briefings/ARCIC%20Overview_211600May09.ppt#256,1,Slide 1

Wolfowitz, Paul, "Meeting Immediate Warfighter Needs (IWNs)," memorandum for Secretaries of the Military Departments, November 15, 2004.

Young, John J., "Defense Science Board (DSB) Task Force on the Fulfillment of Urgent Operational Needs," memorandum for the Chairman, Defense Science Board (DSB), December 17, 2008.

Young, John J., "Defense Science Board Task Force on the Department of Defense Policies and Procedures for the Acquisition of Information Technology," memorandum for Chairman, Defense Science Board, May 1, 2008.